The complete thesaurus consists of
four monolingual volumes
(German, English, French and Italian)
and a quadrilingual index.

The following organizations participated
in the compilation of this thesaurus

MINISTERIE VAN LANDBOUW, LANDBOUWECONOMISCH INSTITUUT, AFDELING DOKUMENTAIRE EN STATISTISCHE ANALYSES, BRUSSEL

FORSCHUNGSGESELLSCHAFT FÜR AGRARPOLITIK UND AGRARSOZIOLOGIE E.V., DOKUMENTATIONSSTELLE, BONN

TECHNISCHE HOCHSCHULE MÜNCHEN, INSTITUT FÜR WIRTSCHAFTSLEHRE DES LANDBAUES, DOKUMENTATIONSSTELLE, WEIHENSTEPHAN

RESEAU DE DOCUMENTATION EN ECONOMIE AGRICOLE (RESEDA), PARIS

MINISTERO DELL'AGRICOLTURA E DELLE FORESTE, DIREZIONE GENERALE DELLA PRODUZIONE AGRICOLA, DIVISIONE INFORMAZIONE E DOCUMENTAZIONE, ROMA

COMMONWEALTH AGRICULTURAL BUREAUX (CAB), COMMONWEALTH BUREAU OF AGRICULTURAL ECONOMICS, OXFORD

under the auspices of

THE COMMISSION OF THE EUROPEAN COMMUNITIES

on the recommendation of the Agricultural Working Group of the Committee
for Information and Documentation in Science and Technology.

Kommission der Europäischen Gemeinschaften
Commission of the European Communities
Commission des Communautés européennes
Commissione delle Comunità europee

Agrarökonomie und ländliche Soziologie Mehrsprachiger Thesaurus

Agricultural Economics and Rural Sociology Multilingual Thesaurus

Economie Agricole et Sociologie Rurale Thesaurus Multilingue

Economia Agraria e Sociologia Rurale Thesaurus Multilingue

2 English Edition

K·G·Saur München·New York·London·Paris 1979

The thesaurus was processed with the ASTUTE software of the Commission of the European Communities and edited by:
Directorate-General for Scientific and Technical Information and Information Management,
Jean Monnet Building, Plateau du Kirchberg,
Luxembourg

EUR 6017

LEGAL NOTICE
Neither the Commission of the European Communities nor any person acting on behalf of the Commission is responsible for the use which might be made of the following information.

K. G. SAUR VERLAG KG
Pössenbacherstr. 2b
D-8000 München 71
Federal Republic of Germany
Tel. (089) 798901
Telex 521 2067 saur d

ISBN 3-598-10099-X (English Edition)
ISBN 3-598-10097-3 (5-vol. Multilingual Set and Index)

CIP-Kurztitelaufnahme der Deutschen Bibliothek

Agrarökonomie und ländliche Soziologie : mehrsprachiger Thesaurus = Agricultural economics and rural sociology = Economie agricole et sociologie rurale / Komm. d. Europ. Gemeinschaften. [Der Thesaurus wurde mit Hilfe d. ASTUTE-Programms d. Komm. d. Europ. Gemeinschaften erstellt u. hrsg. von d. Generaldirektion Wissenschaftl. u. Techn. Information u. Informationsmanagement, Komm. d. Europ. Gemeinschaften. An d. Ausarbeitung dieses Thesaurus waren d. nachstehenden Organisationen beteiligt: Ministerie van Landbouw, Landbouweconom. Inst., Afd. Dokumentaire en Statist. Analyses, Brussel . . .]. — München, New York, London, Paris : Saur.
ISBN 3-598-10097-3

NE: Europäische Gemeinschaften / Kommission; Europäische Gemeinschaften / Generaldirektion Wissenschaftliche und Technische Information und Informationsmanagement; 1. PT; 2. PT

2. English edition. — 1979.
(EUR ; 6017)
ISBN 3-598-10099-X

CLIVE BINGLEY Ltd. &
K. G. SAUR Ltd.
1–19 New Oxford Street
London WC1A 1 NE
United Kingdom
Tel. 01-404 48 18
Telex 24 902 bingle g

K. G. SAUR EDITEUR S.a.r.l.
38, rue de Bassano
F-75008 Paris
France
Tel. 7 23 55-18
Telex Iso Bur. 6 30 144

K. G. SAUR Publishing Inc.
175 Fifth Avenue
New York, N. Y. 10010
USA
Tel. (212) 477-2500
Telex 238 386 kgsp ur

Computer-controlled phototypesetting by
Imprimerie Centrale, Luxembourg
Printed and bound in the Federal Republic of Germany
by Hain-Druck KG, Meisenheim/Glan
for K. G. Saur KG, München

PREFACE

The resolution passed by the Council of Ministers of the European Communities in 1971 defined basic Community policy for the coordination of activities regarding scientific and technical information within the Member States. A first Three-year Community Action Plan in this field was drawn up by the Commission of the European Communities assisted by the Committee for Information and Documentation in Science and Technology (CIDST) and adopted by the Council of Ministers on March 18th 1975. One of its main objectives was to provide for improved methodologies in information technology and, in particular, for the development of suitable multilingual tools to facilitate information exchange.

Three multilingual thesauri covering fields closely related to agriculture have been developed within the framework of this policy under the auspices of the Commission of the European Communities. Directorate General XIII – Scientific and Technical Information and Information Management – and on the recommendation of CIDST. These are: the Veterinary Multilingual Thesaurus, the Food Multilingual Thesaurus and the Agricultural Economics and Rural Sociology Multilingual Thesaurus.

All three have been developed by groups of experts from different professional organizations in Community countries, and each one is quadrilingual – German, English, French and Italian.

Work on the Agricultural Economics and Rural Sociology Multilingual Thesaurus was initiated by a group of experts called by the Agricultural Working Group of CIDST to study the feasibility of establishing a cooperative information system for agricultural management (known as AMIS = Agricultural Management Information System). It was clear that cooperation would involve a preliminary harmonization of indexing practices among the different specialized agencies in the Community through the use of a common multilingual tool. Such a tool would also provide a retrieval aid so that users could search bibliographical data bases using indexing terms in their own language and be able to retrieve references indexed in any of the four languages.

The multilingual thesaurus was compiled over a period of a little more than two years (from the end of 1975 to the end of 1977) and computer processed by the Commission.

With the publication of a first edition of the three thesauri, the Commission hopes to have provided the agricultural information community in Europe with concrete starting points for fruitful cooperation in the three fields covered.

R. K. APPLEYARD
Director General
Scientific and Technical Information and Information Management

CONTENTS

PREFACE V

INTRODUCTION VII

Definition, Scope and Method of Construction VII
Rules and Conventions X
Presentation XV
Participating Organizations, Experts, Bibliographical Sources consulted XVI
Thesaurus Updating XIX

FULL THESAURUS 1

MICROTHESAURI

Agricultural and Rural Law 63
Agricultural Economics 65
Countries 73
Livestock, Crops and Products 77
Methodology 79
R & D and Education 81
Rural Sociology 83

INTRODUCTION

DEFINITION, SCOPE AND METHOD OF CONSTRUCTION

Definition

The ISO Guidelines define a thesaurus, in terms of its function, as a "terminological control device" used to convert the natural language of documents, indexers and users into a structured controlled vocabulary in a specific field for the purpose of indexing and retrieving bibliographic references, and principally those stored as machine-readable data bases. Its major feature is the display of the structure or network of relationships established between the selected terms, and between these (which are the preferred terms for indexing and retrieval and are called "descriptors") and the "non-descriptors" (non-preferred or non-authorized terms which point to the descriptor to be used in their place). This structure defines the semantic content and the use of the term in the given context. The multilingual dimension further links terms through linguistic equivalents in order to generate language versions of the thesaurus. Linguistic equivalents are not merely dictionary translations of terms, for direct translations do not always render the concept which one seeks to represent with a descriptor. Consequently, a thesaurus, whether it is monolingual or multilingual, should neither be considered a comprehensive classification scheme, glossary or dictionary, nor should it be used as such.

Scope

This thesaurus covers the main aspects of agricultural economic policy and management, as well as rural sociology, R & D and extension. Some agricultural law is included in so far as common agreement could be reached on general terms. As a rule, it was not possible at this stage to take specific national legislation and practices into account because they are unique to each country. For more details on the scope of the thesaurus, see the list of microthesauri in the section: Rules and Conventions.

Method of construction

Existing monolingual controlled vocabularies used at the national level were taken as the starting point for collecting terms (see section: Bibliographical Sources Consulted). Selection of terms, introduction of relations and establishment of language equivalents in all four languages of the thesaurus (German, English, French and Italian) were carried out simultaneously in order to take account of national differences in terminology, requirements and approach. ASTUTE, the Commissions's software package for generating, updating, testing, editing and printing mono- and multilingual thesauri was used regularly to process the thesaurus. The programs validated thesaurus data on a formal and a logical level, generated all reciprocal relations and hierarchical structures and printed monolingual and multilingual versions of the thesaurus.

Descriptors were chosen on the basis of their suitability in representing a concept, their impact on indexing and retrieval procedures, current usage in each language and, when known, frequency of use. The main rule observed was that each concept introduced had to be represented by *a descriptor in each language*; no lacunae of linguistic equivalent at descriptor level was allowed, nor could one descriptor in one language be linked voluntarily or inadvertently to two or more different descriptors in one of the other languages (these rules did not apply to non-descriptors and scope notes). Various solutions were applied, as appropriate, to achieve this, and many of these were compromises. This occasionally led to the need to devise solutions (occasionally even to create terms) which can appear artificial in some of the languages in order to accommodate a term judged necessary in one or more of the others. In other cases it sometimes led to the adoption of the foreign language term with an explanation in the scope note. In extreme cases, intractable terms were given the status of non-descriptors in that language, referring to one or more of the accepted descriptors which might fairly adequately represent that particular concept. Such expedients were justified in so far as a thesaurus is already a set of conventions; these were merely additional conventions which permitted the group to come to a workable consensus after analysing concepts.

Examples

(1) "Banque de Travail"

Language	*Chosen Descriptor*	*Explanation*
	BANQUE DE TRAVAIL (Verrechnung von Arbeitszeiten zwischen mehreren Betrieben)	untranslated term with explanatory scope note in German
English	WORK BANKS UF labour banks	artificial, though accepted, term in English, with a non-descriptor
French	BANQUE DE TRAVAIL	technical term in French
Italian	BANCA DI LAVORO	literal translation from the French

The structure in which this term appears further defines its meaning:
it is a narrower term of FARM HELPER SERVICES, which itself is a narrower term of CO-OPERATIVE SERVICES.

(2)The English term "Farm recreations" has been transformed into a non-descriptor:

farm recreations
USE **tourism**

(3)German compound terms with "Recht" (Bodenrecht, Erbrecht...) have been transformed into non-descriptors:

Bodenrecht+
USE **boden**
AND **gesetz**

After completion of a first draft of the thesaurus, it was revised through test indexing on a small scale with each language version as well as careful checking. Amendments involving changes in structure, alterations to language equivalents, corrections, addition or suppression of descriptors, non-descriptors and scope notes were incorporated as a result. Compilation and revision of the thesaurus has taken a little over two years. No thesaurus is ever finished, or ever perfect (and a fortiori, a first edition). The present edition can therefore be considered experimental, on the understanding that it will be subject to revision at regular intervals in order to remain an effective tool. It is indeed expected that the thesaurus will be kept up-to-date and that periodical lists of amendments will be issued, which will be followed by a second edition after a reasonable lapse of time. Persons wishing to remain informed (i.e. receive lists of amendments, transmit suggestions) are requested to complete a form along the lines suggested in the section: Thesaurus Updating.

Compatibility

The three thesauri compiled (see preface) are compatible on a formal level as the same principles have been followed, and the same method of construction and computer processing applied. In terms of their contents, attempts have been made to achieve some measure of compatibility:

– in areas of overlap – the relevant sections of each thesaurus were circulated between groups so that the work of the one could benefit the other. Thus the Food Multilingual Thesaurus was consulted for hierarchies listing fruits, vegetables, food products, meats, food industries; the Veterinary Multilingual Thesaurus was consulted for lists of livestock names and husbandry terms; the list of countries is identical in all three thesauri;
– for common terms – it was sought whenever possible to harmonize descriptors used in more than one thesaurus, at least as far as spelling and language equivalents were concerned.

Nevertheless, complete compatibility between the three was neither entirely feasible, nor considered of highest priority at this stage.

example:

English	*French*	
Brain	**cervelle**	Food Multilingual Thesaurus
	cerveau	Veterinary Multilingual Thesaurus

Each thesaurus had its own coverage and purpose, and it seemed more important that agreement should be reached between participants representing existing national services in the same field, and through them, achieve a measure of compatibility with the national controlled vocabularies in use. For the English language version of the thesaurus, compatibility has been sought in this way with the list of keywords used by the Commonwealth Bureau of Agricultural Economics (C.B.A.E.) for indexing their publications (World Agricultural Economics and Rural Sociology Abstracts; Rural Extension, Education and Training Abstracts; Rural Development Abstracts; Rural Recreation and Tourism Abstracts). In the same way, compatibility has been sought with the RESEDA thesaurus and that of the Ministry of Agriculture for the French language version, and with the „Thesaurus der landwirtschaftlichen Betriebs- und Arbeitswirtschaft" for the German language version.

RULES AND CONVENTIONS

Standards

The following guidelines and standards for monolingual thesauri were observed:

- UNISIST – Guidelines for the establishment and development of monolingual thesauri. UNESCO, SC/WS/555. September 1973.
- ISO 2788 – Documentation – Guidelines for the establishment and development of monolingual thesauri. 1974.
- AFNOR Z 47 100 – Documentation – Règles d'établissement des thésaurus en langue française. Décembre 1973. (Norme expérimentale)
- DIN 1463 – Richtlinien für die Erstellung und Weiterentwicklung deutschsprachiger Thesauri. Oktober 1972. (Vornorm)

Each language version of the thesaurus taken separately is in line with the above-mentioned guidelines. Taken all together, that is, as a single work, the thesaurus remains consistent with these, with the addition of a multilingual dimension. All the languages of the thesaurus were considered equal, although in the course of construction each could in turn become source or target language.

Characteristics

(i) Number of terms in the thesaurus:

Descriptors:	2270 in each language
Non-descriptors:	190 in German
	360 in English
	410 in French
	250 in Italian

(ii) Table of relationships established between thesaurus terms and the abbreviations used to symbolize them:

Type of relationships	*Reciprocal cross-references**	*Symbols***
1 Preferential relations***		
(i) equivalence relations	USE; USE...AND (..AND) USED FOR	USE; USE/AND UF
(ii) alternative relations	SEE...OR.(..OR...) SEEN FOR	SEE/OR (/OR) SF
2 Hierarchical relations	BROADER TERM NARROWER TERM	BT**** NT
3 Associative (or affinitive) relations	RELATED TERM	RT

(iii) Terms (i.e. descriptors and non-descriptors) are arranged in "word blocks" showing, for each alphabetical entry, the full structure: in other words, the network of relationships established between the entry term and the other thesaurus terms. For each descriptor entry, the word block will display the complete hierarchy, that is, all the hierarchical levels of broader and narrower terms up to a maximum of *seven* levels. On the whole, the use of more than 2–4 levels has been avoided wherever possible.

* All relations are reciprocal

** The English symbols have been used uniformly in all language versions to make the thesaurus more homogeneous and also for practical reasons (no standard symbols exist in Italian; French and German standards do not specify symbols for the alternative relation and the notation for this relation could not be more than three characters long). Finally, the English symbols are well known.

*** These relate a non-descriptor to a descriptor (and vice versa); all other relations link descriptors to each other.

**** BT1, BT2 etc. and NT1, NT2 etc. will indicate the hierarchical level of the broader and narrower term for an alphabetical entry (descriptor) and therefore its position in the tree structure.

Examples of word blocks

IMPERFECT COMPETITION	Descriptor entry term
uf *restrictive practices*	Non-descriptor referring to entry term (equivalence relation)
BT1 competition	Broader term to entry term (one level up)
BT2 market economics	Broader term to entry term (two levels up)
NT1 monopoly	Narrower term to entry term (one level down)
NT1 monopsony	Narrower term to entry term (one level down)
NT1	Further narrower terms, including one which is two
NT2	levels down)
rt dumping	Related term

CO-OPERATIVE FARM ENTERPRISES	Descriptor entry term
(co-operation on a single crop or livestock activity)	Scope Note
NT1 crop production group	Narrower terms to entry term (one level down)
⋮	
NT1 grazing group	
⋮	

agronomy	Non-descriptor entry term
USE **agricultural sciences**	Descriptor to which it refers (equivalence relation)

analysis	Non-descriptor entry term
SEE **economic analysis**	Descriptors to which it refers, and amongst which the user must choose (alternative relation)
OR **operations research**	
OR **statistical analysis**	

ECONOMIC ANALYSIS	Descriptor entry term
sf *analysis*	Non-descriptor referring to entry term in reciprocal position (alternative relation)
NT1 marginal analysis	
⋮	

(iv) The thesaurus structure is identical in all language versions with respect to hierarchical and associative relations. Preferential relations (i.e. non-descriptors), as well as scope notes (definitions, explanations, etc.) are however not necessarily identical and depend on the requirements of that language version.

(v) Polyhierarchies, that is the appearance of a descriptor in more than one hierarchy, are permitted provided that the meaning of the descriptor remains the same; its narrower terms, if it has any, will also appear with it in the different hierarchies.

(vi) In general, hierarchies are built on the generic/specific relation between concepts. This principle has however been waived in a number of cases and partitive relations have been included in this thesaurus. Apart from geographical and anatomical descriptors, where such partitive relations between terms are habitual, this applies to descriptors which otherwise would have been isolated and which in the technical literature are usually associated with the descriptor under which they have been subsumed. This also applies to group terms which, in this particular subject area, are difficult to limit to the generic/specific relation. As few descriptors as possible have been left isolated in the structures — that is, without hierarchical relations or, at least, associative relations (related terms) — the thesaurus has only 5 of these.

(vii) Non-descriptors of the USE/AND variety, that is those requiring the combination of *two* or more descriptors to express the concept of the non-descriptor, are followed by a "+" sign so that they can be identified in the reciprocal position under a descriptor entry (with the symbol UF) as involving more than one descriptor, thus avoiding confusion.

Example:

Housewivest+ USE **households** AND **women**	and conversely:	**HOUSEHOLDS** *uf housewivest+* **WOMEN** *uf housewivest+*

(viii) Non-descriptors of the SEE/OR/.. variety (the alternative relation) imply that a choice must be made between the different meanings of the term, and therefore between different descriptors. This relation is particularly useful to resolve homographies and ambiguities. It has also been used where precoordinated descriptors have been preferred to "uniterms", and particularly in English where the adjective precedes the noun (see example given in the word block).

(ix) Related terms have been considered as supplementary indexing and retrieval aids to suggest other descriptors to the user of the thesaurus. Therefore, no special rules have been applied in their establishment. Additional related terms have been introduced however to accommodate the different language versions: they have on occasion been introduced to group descriptors which in one of the languages, because of the alphabetical order in which they appear, are distributed over the whole thesaurus (whereas in one or more of the other languages, they will appear grouped in the alphabetical order). This is often the case in the English language version where the adjective precedes the noun, and this can make it more difficult to find the right descriptor. Related terms are considered part of the structure and therefore appear in all the language versions in the same way.

(x) Scope notes have been used to define descriptors and their use in the thesaurus, explain artificial or untranslated descriptors included to accommodate another language version and, finally, expand any abbreviation used (see examples cited in other sections of this introduction).

(xi) Precoordination (descriptors containing several words) has been introduced where the resulting combination of words represented a single concept frequently used in the technical literature. It has frequently been used to make the meaning of a term more precise to avoid false coordinations in indexing and, on occasion, to express a concept which in one of the other languages might require only one word or be a fixed expression. In other cases, especially in the German language version, precoordinated expressions have been broken down into two or more descriptors. Excessive precoordination (the "pigeon hole" system) was avoided whereever several descriptors were available to express a complex idea without too much ambiguity. Non-descriptors (frequently uniterms referring to one or more precoordinated descriptors) and related terms have been used to link precoordinated descriptors.

(xii) For practical reasons, and in particular for computer processing with the ASTUTE software package, the length of descriptors has been limited to a maximum of 35 characters. In the majority of cases this has been sufficient to formulate descriptors and non-descriptors adequately; in a few cases abbreviations were unavoidable (see below). This restriction does not apply to scope notes which can use as many lines as are required.

Term control

(i) *Word order:* descriptors, particularly precoordinated ones, appear in the natural word order. Inverted forms have on occasion, though rarely, been used in non-descriptors. Qualifiers between brackets forming part of a descriptor have only been used exceptionally, for the subdivision into regions of Community countries (see below: special options).

(ii) *Alphabetical order:* since the thesaurus has been processed by computer, the alphabetical sequence of the entries is based on a character by character sort.

(iii) *Singular and plural forms:* plural forms have generally been adopted for the English language version. The French language version has however systematically used singular forms in accordance with the AFNOR standard, except in special cases where it was necessary to distinguish between two descriptors with widely different meanings.

Example from the French language version:

Culture = intellectual
Cultures = agricultural

(note : the same may occur in the other language versions!)

The German and Italian versions have on the whole followed suit with singular forms.

(iv) *Abbreviations :* they have as far as possible been avoided except for the names of organizations; these have systematically been abbreviated because the full form of some names required more than 35 characters. The limitation to 35 characters has moreover led to a few other (unavoidable) abbreviations. In all cases, all abbreviations have been expanded in scope notes.

Examples :

EFTA
(european free trade association)
GAEC
(groupement agricole
d'exploitation en commun)
APPLICAZIONE E VALUTAZIONE DELLA PI
(applicazione e valutazione della pianificazione)
AKP-LAENDER
(assoziierte laender afrikas,
der karibischen see und des pazifiks)

In the German language version, the following adjectives have been systematically abbreviated in each occurrence :

- LANDW (and compounds INNERLANDW, NICHTLANDW) = landwirtschaftlich ..
- WIRTSCHAFTL = wirtschaftlich ..
- AUSSERBETRIEBL (and other compounds EINZELBETRIEBL) = ausserbetrieblich ..

and a further two abbreviations with names of regions in the list of countries (See section : Special Options).

(v) *Punctuation, diacritics and accents :* the character set used to process the thesaurus does not permit the use of accents and diacritics. These have therefore been omitted in all languages. Following current practice, the German umlaut has been converted to ae, oe, ue. Furthermore, to facilitate the use of the thesaurus in machine-readable form, all punctuation has been removed as far as possible, including the full stop in abbreviations and the "accento" in Italian. The hyphen has been treated somewhat differently : in the French language version, all hyphens have been eliminated; they have however occasionally been maintained in the English, German and Italian language versions either where they act as a link in an artificially formed expression, or where they are considered habitual.

Examples :

LAND-STADT-WANDERUNG
BOOK-KEEPING SYSTEMS
COMPLESSO AGRO-INDUSTRIALE

Coverage

The thesaurus has been subdivided into seven conceptual fields or 'Microthesauri' for the sake of convenience during the compilation period and to facilitate its use. These are loose groupings of complete hierarchies; hierarchies can appear in several microthesauri if necessary; some microthesauri contain more descriptors than others.

The microthesauri are the following :

- Agricultural and Rural Law
- Agricultural Economics
- Countries
- Livestock, Crops and Products
- Methodology

- R&D and Education
- Rural Sociology

Special Options

(i) In agreement with UNESCO, the structured list of countries has been taken over from the SPINES thesaurus (International System for the Exchange of Information on Science and Technology for Policy-Making, Management and Development) after updating. This solution was chosen so that an identical list could be used for all three thesauri and have an international standing. Since the SPINES thesaurus was only available in English at the time of compilation, the other three language versions were drawn from the ILO thesaurus, the EUDISED thesaurus and the nomenclature of countries used by the Statistical Office of the European Communities (OJ n° L356, 28 December 1976).

A supplementary subdivision into regions for the nine countries of the European Communities has been included in this thesaurus. In order to simplify the list of regions at this stage, it has been limited to the minimum requirements of each country for national use and for information exchange, i.e. to the general political/administrative regions (e.g. the 'Länder' of the Federal Republic of Germany, the main regions in France). Further subdivision (e.g. into agricultural regions) will be included when the thesaurus is updated, if the need is established. For the present, no attempt has been made to translate the proper names of the regions in the various language versions, with the exception of English where the current practices of C.B.A.E. (Commonwealth Bureau of Agricultural Economics) were observed because of the wider scope of their services. The basic premise adopted in the other language versions was that French, Belgian, German and Italian indexers would only use the regional terms to index items from their own countries, and obtain the remainder through exchanges within the information network. For this reason, as well as to facilitate alphabetical grouping and up-posting, the names of regions are combined with the name of the country and appear in brackets. The conventions requested for each language are respected.

Examples :

FRANKREICH (R. BRETAGNE)
ROYAUME UNI (W.MIDLANDS)
PAESI BASSI (ZEELAND)

In the German language version, two abbreviations have been used in conjunction with names of regions because of the limitation to 35 characters :

GB+ name of the region, in the place of VEREINIGTES KOENIGREICH
BRD+ name of the Länder, in the place of BUNDESREPUBLIK DEUTSCHLAND

In the French language version, the "départements" appear as non-descriptors linked to the relevant regions.

(ii) The microthesaurus for Agricultural and Rural Law has been limited to purely legal terms. Other terms which can be used in conjunction with these legal terms are to be found elsewhere in the thesaurus.

(iii) Rural sociology has been covered pragmatically. The microthesaurus dealing with this topic is not intended to be a theoretical survey of the subject.

PRESENTATION

The complete thesaurus is presented in *four* monolingual volumes and *one* multilingual index volume.

Each monolingual volume contains:

- the full thesaurus in the relevant language, sorted alphabetically on all descriptors and non-descriptors and displaying, for each entry, the relational structure in a "word block" (see above: Rules and Conventions).
- a systematic section, in the relevant language, presented by microthesauri. Within each microthesaurus, descriptors are arrayed in complete hierarchies (or tree-structures) only, starting from the most generic descriptor and going down to the most specific one. Individual descriptors therefore do not have separate entries in alphabetical order, except for the most generic one of the hierarchy. Non-descriptors do not appear, nor do related terms between individual descriptors. This systematic section is included to permit a rapid overview of the thesaurus coverage.

The index volume contains the linguistic equivalents of all descriptors, arranged in quadrilingual alphabetical lists sorted successively on each of the four languages. The structure is not displayed. Microfiches displaying descriptors in trilingual structured arrays are located in a pocket at the back of the index volume. The microfiches are only included for reference, on occasions when the user might need to check the parallel structure of the thesaurus in several languages.

Ariane I. ILJON
Responsible for coordination
Directorate General Scientific and Technical Information
and Information Management

PARTICIPATING ORGANIZATIONS, EXPERTS, BIBLIOGRAPHICAL SOURCES CONSULTED

Participating Organizations

- Ministerie van Landbouw, Landbouweconomisch Instituut, Afdeling Dokumentaire en Statistische Analyses, Brussel
- Forschungsgesellschaft für Agrarpolitik und Agrarsoziologie e.V., Dokumentationsstelle, Bonn
- Technische Hochschule München, Institut für Wirtschaftslehre des Landbaues, Dokumentationsstelle, Weihenstephan
- Réseau de Documentation en Economie Agricole (RESEDA), Paris
- Ministero dell'Agricoltura e delle Foreste, Direzione Generale della Produzione Agricola, Divisione Informazione et Documentazione, Roma
- Commonwealth Agricultural Bureaux (CAB), Commonwealth Bureau of Agricultural Economics, Oxford

Experts

Members of the working group who compiled the thesaurus:

ASCANI, F., Ministero dell'Agricoltura e delle Foreste, Roma
GOEDSEELS, F., Landboueconomisch Instituut, Brussel
KESTNER, J., Commonwealth Bureau of Agricultural Economics, Oxford
LUNZE, J., Dokumentationsstelle für Agrarpolitik, landwirtschaftliches Marktwesen und ländliche Soziologie, Bonn
LUPORI, N., ministero dell'Agricoltura e delle Foreste, Roma (Consultant)
PUJES, P., RESEDA, Paris
RAHMANN, M., Dokumentationsstelle für landw. Betriebs- und Arbeitswirtschaft, Weihenstephan

Experts, whose support has made the enterprise possible:

D'AMBROSIO, A., Ministero dell'Agricoltura e delle Foreste, Roma
DIEUDONNE, D., RESEDA, Paris
JONES, O., Commonwealth Bureau of Agricultural Economics, Oxford

and, amongst the many who have contributed, were:

DEVISCH, N., Landbouwingenieur, Leuven
LATTWEIN, J. N., Ministère de l'Agriculture, Paris
MATTEI, J. N., Ministère de l'Agriculture, Paris
ROMAGNOLI, E., Ministero dell'Agricoltura e delle Foreste, Roma (Consultant)
THIRKELL SMITH, A., Commonwealth Bureau of Agricultural Economics, Oxford
TROISIER, M., INRA, Paris
VERTESSEN, J., Landbouwingenieur, Leuven

Coordination:

ILJON, A. I., Commission of the European Communities, Directorate General XIII — Scientific and Technical Information and Information Management, Luxembourg

Bibliographical Sources consulted

Primary sources:

Caisse Nationale de Crédit Agricole: Thesaurus agricole, économique, financier et juridique. 3e édition. (Paris, FR; C.N.C.A., 1974).

Commonwealth Agricultural Bureaux: Consolidated list of Keywords used in all CAB abstracting journals. (Farnham Royal, GB; unpublished, 1977).

Commonwealth Bureau of Agricultural Economics: List of Keywords used for indexing World Agricultural Economics and Rural Sociology Abstracts. (Oxford, GB; unpublished, 1977).

Dokumentationsstelle für Agrarpolitik, landwirtschaftliches Marktwesen und ländliche Soziologie: Liste der Deskriptoren. (Bonn, DE; unveröffentlicht).

Rahmann, M.: Thesaurus der landwirtschaftlichen Betriebs- und Arbeitswirtschaft. 4. Auflage. (Freising-Weihenstephan, DE; Dokumentationsstelle für Landwirtschaftliche Betriebs- und Arbeitswirtschaft, 1968).

Réseau de Documentation en Economie Agricole: Thesaurus RESEDA "Economie Agricole". Edition provisoire. (Paris, FR).

Viet, J.: Thesaurus for information processing in sociology, Thesaurus pour le traitement de l'information en sociologie. (Paris, FR; the Hague, NL; Mouton, 1971).

Organisation de Coopération et de Développement Economiques: Macrothesaurus. Liste des principaux descripteurs relatifs au développement économique et social. Edition française. (Paris, FR; OCDE, 1974).

Secondary sources: (incl. dictionaries)

Bannock, G., Baxter, R. R., Rees, R.: Dizionario di Economia. (Bari, IT; Editori Laterza, 1977).

Bara, L.: Dizionario di Politica Economica. (Roma, IT; Editori Riuniti, 1974).

Belgische Dienst voor Buitenlandse Handel. Nat. Centrum voor Buitenlandse Handel: Thesaurus. (Brussel, BE; unpublished).

Bernsdorf, W., Hrsg.: Wörterbuch der Soziologie. (Stuttgart, DE; Enke Verlag, 1969).

Brandes, W.: Wie analysiere und plane ich meinen Betrieb? 2. Aufl. (Hamburg–Berlin, DE; Paul Parey, 1974).

Brandes, W., Woermann, E.: Landwirtschaftliche Betriebslehre – Theorie und Planung des landwirtschaftlichen Betriebes. Band 1, Allgemeiner Teil. (Hamburg–Berlin, DE; Paul Parey, 1969).

Brandes, W., Woermann, E.: Landwirtschaftliche Betriebslehre – Organisation und Führung landwirtschaftlicher Betriebe. Band II, Spezieller Teil. (Hamburg–Berlin, DE; Paul Parey, 1971).

Brunoli, A.: Dizionario Avicolo Internazionale. (Bologna, IT; Edagricole, 1967).

Commission of the European Communities, Translation division: European Communities glossary. French/English. 6th Edition. (Brussels, BE; Roneotyped, 1975).

Conte, A.: Dizionario Giuridico et Economico. (Milano, IT; Giuffre', vol 1 1969, vol 2 1971).

Deutsche Stiftung für Internationale Entwicklung: Thesaurus für wirtschaftliche und soziale Entwicklung. 7. rev. Auflage. (Hamburg–Bonn, DE; Deutsche Stiftung für Internationale Entwicklung, 1977).

Eichborn, R. von: Wirtschaftswörterbuch, I. Englisch/Deutsch, II. Deutsch/Englisch. (Düsseldorf, DE; Econ-Verlag, 1965).

Endruweit, G.: Dreisprachiges Wörterbuch der Soziologie. (Neuwied, DE; Hain Verlag, 1975).

European Parliament, Translation division: Economic Terminology. 2nd supplement. (Luxembourg, LU; Roneotyped, 1973).

Favati, F.: Dizionario di Agricoltura. (Bologna, IT; Edagricole, 1971).

Gunston, D. A., Corner, V. M.: Deutsch-Englisches Glossarium finanzieller und wirtschaftlicher Fachausdrücke. (Frankfurt/Main, DE; Fritz Knapps Verlag, 1967).

Haensch, G., Haberkamp de Anton, G., editors: Dictionary of Agriculture. English, German, French, Russian. (Amsterdam, NL; Oxford, GB; New York, US; Elsevier Scientific publishing company, 1975). 1000 pp., and earlier editions BLV Verlagsgesellschaft, 1966, 1969.

Haensch, G., Haberkamp de Anton, G., Leninger, G., editors: Dizionario di Agricoltura. Supplemento Italiano. (München–Basel–Wien, DE, CH, AT; BLV Verlagsgesellschaft, 1963).

HLBS: Begriffs-Systematik für die landwirtschaftliche und gartenbauliche Betriebslehre. H. 14, 5. Aufl. (Bonn, DE; Schriftenreihe des HLBS, 1973).
Hofmann, E.: Wörterbuch der Datenverarbeitung. Englisch/Deutsch, Deutsch/Englisch. (Berlin, DD; Verlag der Wirtschaft, 1971).
IGORB: Internationale Große Operations Research Bibliographie, vol 1, No. 1. (Darmstadt, DE; Kontohow's I&D Service, 1976).
International Institute for Legal and Administrative Terminology: European glossary of legal and administrative terminology. Vol 18. Regional policy, German/English Glossary. Europa-Glossar der Rechts- und Verwaltungssprache, Band 18. Raumordnung. Deutsch/Englisch. (Berlin, DE; London, GB; Langenscheidt, Sweet & Maxwell, 1975).
Landbouweconomische Instituut: Thesaurus (gecontroleerde wordenlijst). (Brussel, BE, unpublished).
Leatherdale, D.: Canadian Agricultural Thesaurus. 4 Vols. (Ottawa, CA; Research Branch, Canada Dept. of Agriculture, 1971).
Meimberg, P.: Landwirtschaftliches Rechnungswesen. (Stuttgart, DE; Eugen Ulmer Verlag, 1966).
Merchi, F. de, Elena, A.: Dizionario di Sociologia. (Roma, IT; Edizioni Paoline, 1976).
Minstry of Agriculture, Fisheries and Food: Terms and definitions used in farm and horticultural management. (London, GB; Ministry of Agriculture, Fisheries and Food, 1970).
Morelli, M.: Dizionario di Informatica e degli Elaboratori Elettronici. (Milano, IT; Franco Angeli Editore, 1975).
Morice, E.: Dizionario di Statistica. (Milano, IT; Isedi, 1971).
Musmara, A.: Dizionario Botanico. (Bologna, IT; Edagricole, 1972).
Napoleoni, C.: Dizionario di Economia Politica. (Milano, IT; Edizioni comunita, 1956).
National Board for Prices and Incomes: Prices, profits and costs in food distribution. Report No. 165. Appendix B. Glossary of terms. (London, GB; National Board for Prices and incomes, 1971).
Newhouse, J. W. A.: A handbook of agricultural (arable farming) terms. English/French. (Bishops Waltham, Hampshire, GB; self published, 1972).
Newhouse, J. W. A.: A handbook of livestock rearing, feeding and marketing terms. English/French. (Bishops Waltham, Hampshire, GB; self published, 1973).
Papi, N.: Dizionario di Economia. (Torino, IT; Utet, 1971).
Piccioni, M.: Dizionario per gli Alimenti del Bestiame. (Bologna, IT; Edagricole, 1968).
Reisch, E., Zeddis, J.: Einführung in die landwirtschaftliche Betriebslehre. Band 2. Spezieller Teil – Grundlagen und Methoden der Entscheidung, Ökonomik der pflanzlichen und tierischen Produktion. (Stuttgart, DE; Uni-Taschenbüdher, Band 617, 1977).
Stahn, E.: Glossarium entwicklungspolitischer Ausdrücke. Deutsch, Englisch, Französisch. (Baden-Baden, DE; Nomoss Verlag, 1969).
Statistical Office of the European Community: Yearbook of agricultural statistics. Glossary of principal terms used. (Luxembourg, LU; CEC, 1976).
Steinhauser, H., Lanbahn, C., Peters, U.: Einführung in die landwirtschaftliche Betriebslehre. Band 1: Allgemeiner Teil – Produktionsgrundlagen, Produktionstheorie und Rechnungssysteme mit Planungsrechnungen. (Stuttgart, DE; Uni-Taschenbücher, Band 113, 1972).
Steinmetz, H., Lohmann & CoKG: Gartenbautechnik I. Mehrsprachen-Bildwörterbuch. Deutsch, English, Français, Espagnol, Italiano, Nederlands. (Betzdorf/Sieg, DE; H. Steinmetz, 1964). Tecnica ortofruticola e di giardinaggio. Dizionario poliglotta illustrato. (Bologna, IT; Edagricole, 1973).
Steinmetz, H., Lohmann & CoKG: Landmaschinen und Geräte. Mehrsprachen-Bildwörterbuch. Deutsch, English, Français, Espagnol, Italiano, Nederlands. (Betzdorf/Sieg, DE; H. Steinmetz, 1972). Macchine e attrezzi agricoli. Dizionario poliglotta illustrato. (Bologna, IT; Edagricole, 1973).
Steinmetz, H., Lohmann & CoKG: Tierfütterung und Tierhaltung. Mehrsprachen-Bildwörterbuch. Deutsch, English, Français, Espagnol, Italiano, Nederlands. (Betzdorf/Sieg, DE; H. Steinmetz, 1966). Alimentazione e governo del bestiame. Dizionario poliglotta illustrato. (Bologna, IT; Edagricole, 1973).
Stephenson, I.: The law relating to agriculture. (Farnborough, GB; Saxon House, 1975).
Theodorson, G. A., Theodorson, A. G.: Dizionario di Sociologia. (Milano, IT; Alberto Marotta Editore, 1975).

UNESCO: Spines thesaurus. A controlled and structured vocabulary of Science and Technology for Policy-Making, Management and Development. (Paris, FR; the Unesco Press, 1976). 80 pp, 316 pp, 246 pp, 34 Tables.

University of Reading, Agricultural Extension and Rural Development Centre: Thesaurus of terms for rural extension, revised edition. (Reading, GB; roneotyped, 1974).

US Agency for International Development (AID): Food production and nutrition, research literature for development. Vol 1. Non geographical descriptors. (Washington D.C., US; US AID, 1976).

USA Department of Agriculture: Vocabulary of the American Agricultural Economics Documentation Center. (unpublished, 1970).

Werner, H., König, I.: Glossare zur Arbeitsmarkt- und Berufsforschung. Deutsch/Englisch. (Stuttgart, DE; Verlag W. Kohlhammer, 1974).

–,: Handwörterbuch der Sozialwissenschaften. 12 Bände. (Stuttgart, DE; Fischer Verlag 1965).

–,: Practical definitions and formulae. In: Economist Measurement Guide. (London, GB; Economist Newspapers Ltd, 1975).

THESAURUS UPDATING

The thesaurus will be regularly updated.

A. Suggestions for alterations to the thesaurus are welcome. If you would like to submit amendments, corrections or additions for consideration by the compilers of the thesaurus, please do so using the following format and with one suggestion per sheet:

1. Language version affected:
2. Proposed amendment/correction/new term:
3. Brief justification or explanation:
4. Your name:
5. Your institution + address:
6. Date of submission:

Remember that amendments, corrections and additions may affect all language versions. Send your suggestions to the address below.

B. If you would like to be put on the mailing list to receive notices of modifications, please write to the address below.

AGRICULTURAL ECONOMICS AND RURAL SOCIOLOGY
MULTILINGUAL THESAURUS
Commission of the European Communities,
Directorate-General XIII/A–3,
Bâtiment Jean Monnet,
Plateau du Kirchberg,
LUXEMBOURG (Grand Duché)

FULL THESAURUS

ABANDONED LAND
BT1 land

ABATTOIRS
uf *slaughterhouses*
BT1 meat and livestock industry
BT2 food industry

ABSTRACTING
BT1 document analysis
BT2 information processing

abyssinia
USE **ethiopia**

ACARICIDES
BT1 pesticides

ACCESSION
(landlord's acquisition of buildings erected by tenant)
BT1 acquisition of ownership
rt reclaimed land

ACCIDENT BENEFITS
BT1 social security
rt insurances

ACCIDENT PREVENTION
uf *accidents at work*
BT1 safety at work
BT2 working conditions
rt invalid benefits

accidents at work
USE **accident prevention**

ACCOUNTING
NT1 aggregate farm accounts
rt aggregate accounts
rt balance sheet analysis
rt book-keeping
rt cost analysis
rt farm planning
rt farm results
rt income
rt input-output analysis
rt national accounting
rt regional accounting

ACCOUNTING BALANCE
BT1 balance sheet analysis

ACCULTURATION
BT1 cultural environment
BT2 culture

ACP
(african caribbean pacific partners to the lome convention with the eec)
BT1 organizations

ACQUISITION OF OWNERSHIP
NT1 accession
NT1 contracts
NT1 possession
NT1 property protection
NT1 succession
NT2 indemnification
NT2 indivisible estates
NT2 partition of estates

acreage
USE **farm area**

ACREAGE SUBSIDIES
BT1 subsidies
BT2 agricultural financial policy
rt structural policy
rt support measures

ACTION
BT1 decision making

ACTIVATING PRICES
(cap price for wine)
BT1 prices

acts of parliament
USE **laws**

ACTUAL COSTS
BT1 production costs
BT2 costs
BT3 cost analysis

ACTUARIAL MATHEMATICS
BT1 financial mathematics
BT2 mathematics

ADDITIVES
NT1 antibiotics
NT1 colourings
NT1 flavourings
NT1 mineral supplements
NT1 preservatives
NT1 vitamins
rt feedingstuffs
rt food
rt manufactured feeds
rt preservation

aden
USE **yemen pdr**

ADJUSTMENT OF PRODUCTION
BT1 production policy
NT1 agricultural adjustment
NT1 dairy-beef conversion schemes
rt setting-up grants

ADMINISTRATION
NT1 central government
NT2 ministries of agriculture
NT1 local government
NT2 local authority areas
NT1 private organizations
NT1 public authorities
NT1 regional government
NT1 semi-private organizations

ADMINISTRATIVE AREAS
BT1 regions

ADOPTION OF INNOVATIONS
BT1 modernization
BT2 social change

ADULTS
BT1 age

ADVERSE POSSESSION
BT1 ownership limitation
BT2 ownership

ADVERTIZING
BT1 publicity
BT2 sales promotion
BT3 marketing techniques

ADVISORY CENTRES
BT1 extension
rt professional services

ADVISORY OFFICERS
BT1 extension

AERIAL SURVEYS
BT1 agricultural aviation
rt satellite surveys
rt topography

AFARS AND ISSAS TERRITORY
uf *french somaliland*
BT1 east africa
BT2 africa south of sahara
BT3 africa
rt france
rt french-speaking africa

AFFORESTATION
uf *re-afforestation*
BT1 forest management
BT2 land use systems

AFGHANISTAN
BT1 west asia
BT2 asia

AFRICA
NT1 africa south of sahara
NT2 central africa
NT3 burundi
NT3 cameroon
NT3 central african empire
NT3 chad
NT3 congo
NT3 equatorial guinea
NT3 gabon
NT3 rwanda
NT3 sao tome and principe
NT3 zaire
NT2 east africa
NT3 afars and issas territory
NT3 burundi
NT3 ethiopia
NT3 kenya
NT3 malawi
NT3 rwanda
NT3 seychelles
NT3 somalia
NT3 sudan
NT3 tanzania
NT3 uganda
NT2 southern africa
NT3 angola
NT3 botswana
NT3 comoro islands
NT3 lesotho
NT3 madagascar
NT3 malawi
NT3 mauritius
NT3 mozambique
NT3 namibia
NT3 reunion
NT3 rhodesia
NT3 south africa
NT3 st helena
NT3 swaziland
NT3 zambia
NT2 west africa
NT3 benin
NT3 cameroon
NT3 cape verde
NT3 gambia
NT3 ghana
NT3 guinea
NT3 guinea-bissau
NT3 ivory coast
NT3 liberia
NT3 mali
NT3 mauritania
NT3 niger
NT3 nigeria
NT3 senegal
NT3 sierra leone
NT3 st helena
NT3 togo
NT3 upper volta
NT3 western sahara
NT1 north africa
NT2 algeria
NT2 egypt
NT2 libya
NT2 morocco
NT2 tunisia
rt english-speaking africa
rt french-speaking africa

africa (maghreb)
SEE **algeria**
OR **morocco**
OR **tunisia**

AFRICA SOUTH OF SAHARA
BT1 africa
NT1 central africa
NT2 burundi
NT2 cameroon
NT2 central african empire
NT2 chad
NT2 congo
NT2 equatorial guinea

AFRICA SOUTH OF SAHARA (C)
NT2 gabon
NT2 rwanda
NT2 sao tome and principe
NT2 zaire
NT1 east africa
NT2 afars and issas territory
NT2 burundi
NT2 ethiopia
NT2 kenya
NT2 malawi
NT2 rwanda
NT2 seychelles
NT2 somalia
NT2 sudan
NT2 tanzania
NT2 uganda
NT1 southern africa
NT2 angola
NT2 botswana
NT2 comoro islands
NT2 lesotho
NT2 madagascar
NT2 malawi
NT2 mauritius
NT2 mozambique
NT2 namibia
NT2 reunion
NT2 rhodesia
NT2 south africa
NT2 st helena
NT2 swaziland
NT2 zambia
NT1 west africa
NT2 benin
NT2 cameroon
NT2 cape verde
NT2 gambia
NT2 ghana
NT2 guinea
NT2 guinea-bissau
NT2 ivory coast
NT2 liberia
NT2 mali
NT2 mauritania
NT2 niger
NT2 nigeria
NT2 senegal
NT2 sierra leone
NT2 st helena
NT2 togo
NT2 upper volta
NT2 western sahara

AGE
NT1 adults
NT1 age differences
NT1 age groups
NT1 age structure
NT2 population pyramids
NT1 childhood
NT1 generations
NT1 old age
NT1 youth
NT2 rural youth

AGE DIFFERENCES
BT1 age

AGE GROUPS
BT1 age
BT1 groups

AGE STRUCTURE
BT1 age
NT1 population pyramids

AGGREGATE ACCOUNTS
NT1 national accounting
NT1 regional accounting
rt accounting

AGGREGATE DATA
NT1 consumer expenditure
NT2 household expenditure
NT1 final consumption
NT1 final production
NT1 gross national product

AGGREGATE DATA (C)
NT1 household consumption
NT1 intermediate consumption
NT1 national expenditure
NT1 total agricultural expenditure
NT1 value added
rt economic accounts

AGGREGATE FARM ACCOUNTS
BT1 accounting

AGRARIAN COUNTRIES
BT1 economic geography

AGRARIAN REFORM
NT1 land reform
rt economic development
rt structural policy

AGRIBUSINESS
BT1 agro-industrial sector
rt vertical integration

AGRICULTURAL ADJUSTMENT
(includes horticulture)
BT1 adjustment of production
BT2 production policy

AGRICULTURAL AVIATION
sf *aviation*
NT1 aerial surveys
rt air transport

AGRICULTURAL BANKS
uf *land mortgage banks*
BT1 banks

AGRICULTURAL BILLS OF EXCHANGE
rt credit transactions

AGRICULTURAL BUDGET
BT1 public finance

AGRICULTURAL CENSUSES
BT1 censuses

AGRICULTURAL CREDIT
uf *supervised credit+*
BT1 credit
BT2 credit transactions
rt co-operative credit

AGRICULTURAL CRISIS
BT1 agricultural development
rt economic depression

AGRICULTURAL DEBENTURES
(privileged debts for farm equipment)
BT1 debt
BT2 credit transactions
rt credit

AGRICULTURAL DEVELOPMENT
NT1 agricultural crisis
NT1 green revolution
NT1 rural-urban migration
rt agricultural situation
rt sectoral development

AGRICULTURAL DISASTERS
NT1 fires
NT2 forest fires
NT1 flooding
NT1 tempest
rt climatic factors
rt compensation
rt crop insurance
rt damage
rt natural disasters
rt subsidies

AGRICULTURAL ECONOMICS
BT1 economics

AGRICULTURAL ECONOMISTS
BT1 research workers
BT2 research personnel
BT3 organization of research

AGRICULTURAL EDUCATION
BT1 education

AGRICULTURAL ENGINEERING
rt land improvement
rt mechanization
rt professional services
rt technical progress

AGRICULTURAL FINANCIAL POLICY
NT1 subsidies
NT2 acreage subsidies
NT2 grubbing up grants
NT2 improvement grants
NT2 non-marketing premiums
NT2 setting-up grants
NT2 severance allowance
NT2 slaughter premiums
NT2 stock-piling subsidies
rt agricultural policy
rt policy

AGRICULTURAL GEOGRAPHY
BT1 rural environment
BT2 environment
rt geography
rt rural areas
rt social geography

AGRICULTURAL GUIDELINES
BT1 production controls
BT2 production policy

AGRICULTURAL HOUSEHOLDS
BT1 households
BT2 economic units

AGRICULTURAL INSURANCE
BT1 insurances
NT1 crop insurance
NT1 fire insurance
NT1 hail insurance
NT1 livestock insurance

AGRICULTURAL LAW
rt contract legislation

AGRICULTURAL MACHINERY INDUSTRY
BT1 input industries

AGRICULTURAL MANPOWER
NT1 child labour
NT1 elderly family-workers
NT1 family labour
NT1 hired labour
NT2 female labour
NT2 migrant labour
NT2 part time labour
NT2 permanent labour
NT2 seasonal labour
NT2 skilled labour
NT2 unskilled labour
NT1 manpower units
NT1 young workers
rt agricultural work force
rt farm workers
rt labour
rt labour economics
rt working population

AGRICULTURAL PLANNING
BT1 planning
rt farm planning

AGRICULTURAL POLICY
NT1 cap
NT1 marketing policy
NT2 market regulations
NT3 intervention
NT2 market stabilization
NT2 price policy
NT3 compensatory amounts
NT1 national agricultural policy
NT1 regional agricultural policy
NT2 less favoured areas
NT3 mountain areas
NT1 structural policy
NT2 land consolidation

AGRICULTURAL POLICY (C)
NT1 support measures
rt agricultural financial policy
rt agricultural trade
rt food policy
rt forest policy
rt land diversion
rt policy
rt production policy

AGRICULTURAL POPULATION
BT1 population structure
BT2 population

AGRICULTURAL PRESS
BT1 newspapers
BT2 mass media
rt extension

AGRICULTURAL PRICES
BT1 prices
NT1 guaranteed minimum producer prices
NT1 guide prices
NT1 intervention prices
rt price policy
rt producer prices
rt target prices

AGRICULTURAL PRODUCTS
BT1 products

AGRICULTURAL REGIONS
uf zones
BT1 regions

AGRICULTURAL RESEARCH
BT1 research
rt agricultural sciences

AGRICULTURAL SCIENCES
uf agronomy
rt agricultural research

agricultural sector
SEE **agro-industrial sector**
OR **farm sector**

AGRICULTURAL SITUATION
NT1 capital allocation
NT1 disparity
NT1 farm indebtedness
NT2 bankruptcy
NT1 investment
NT2 foreign investment
NT2 private investment
NT2 public investment
NT2 reinvestment
NT1 parity
NT1 productivity
NT2 capital productivity
NT2 labour productivity
NT2 land productivity
NT1 resource allocation
rt agricultural development
rt economic analysis

AGRICULTURAL SOCIETY
BT1 societies
NT1 peasantry

AGRICULTURAL STRUCTURE
NT1 farm structure
NT1 production structure
rt structure

AGRICULTURAL SYSTEMS
NT1 commercial farming
NT1 extensive farming
NT1 intensive farming
NT1 subsistence farming
NT1 transitional farming
NT1 types of farming
NT2 collective farming
NT2 domestic gardens
NT2 family farms
NT2 full-time farming
NT2 large farms
NT2 medium sized farms

AGRICULTURAL SYSTEMS (C)
NT2 mixed farming
NT2 part-time farming
NT3 peasant workers
NT2 plantations
NT2 small farms
NT2 specialized farming
rt farm structure

AGRICULTURAL TENURE
NT1 farm leases
NT1 grazing tenancy
NT1 heritable leases
NT1 livestock leases
rt landowners
rt leases
rt period of tenancy
rt resumption of possession
rt tenants

AGRICULTURAL TRADE
BT1 external trade
BT2 trade
rt agricultural policy
rt trade

AGRICULTURAL TRADE UNIONS
BT1 interest groups
BT2 groups
rt trade unions

AGRICULTURAL TRIBUNALS
rt jurisprudence

AGRICULTURAL WASTES
BT1 pollution by agriculture
BT2 environmental policy
NT1 crop residues
NT1 livestock wastes
rt waste disposal

AGRICULTURAL WORK FORCE
BT1 working population
BT2 population structure
BT3 population
rt agricultural manpower
rt labour economics

AGRO-INDUSTRIAL RELATIONS
uf industry and agriculture
BT1 production controls
BT2 production policy

AGRO-INDUSTRIAL SECTOR
uf food and agricultural sector
sf agricultural sector
NT1 agribusiness
NT1 post-agricultural sector
NT1 pre-agricultural sector

agronomy
USE **agricultural sciences**

aid
SEE **development aid**
OR **food aid**
OR **technical aid**

AIR POLLUTION
BT1 pollution of agriculture
BT2 environmental policy

AIR TRANSPORT
sf aviation
BT1 transport
rt agricultural aviation

ALBANIA
BT1 eastern europe
BT2 europe
BT1 mediterranean countries

ALCOHOL
uf ethyl acohol
rt alcoholic beverages

ALCOHOLIC BEVERAGES
BT1 beverages

ALCOHOLIC BEVERAGES (C)
NT1 beer
NT1 cider
NT1 liqueurs
NT1 spirits
NT1 wine
NT2 champagne and sparkling wines
NT2 quality wine
NT2 table wine
rt alcohol

ALCOHOLISM
BT1 deviant behaviour
BT2 social crisis
BT3 socialization

ALGERIA
sf africa (maghreb)
BT1 arab countries
BT1 mediterranean countries
BT1 north africa
BT2 africa
rt french-speaking africa

allotments
USE **domestic gardens**

ALMONDS
BT1 nuts
NT1 bitter almonds
NT1 sweet almonds

ALTAI PROGRAMME
BT1 project control
BT2 operations research

AMERICA
NT1 caribbean
NT2 bahamas
NT2 barbados
NT2 cayman islands
NT2 cuba
NT2 dominican republic
NT2 guadeloupe
NT2 haiti
NT2 jamaica
NT2 martinique
NT2 montserrat
NT2 netherlands antilles
NT2 puerto rico
NT2 tortola
NT2 trinidad and tobago
NT2 turks and caicos islands
NT2 united states virgin islands
NT2 west indies associated states
NT3 antigua
NT3 dominica
NT3 grenada
NT3 st kitts and nevis and anguilla
NT3 st lucia
NT3 st vincent
NT1 central america
NT2 belize
NT2 costa rica
NT2 cuba
NT2 dominican republic
NT2 el salvador
NT2 guatemala
NT2 haiti
NT2 honduras
NT2 jamaica
NT2 nicaragua
NT2 panama
NT2 panama canal zone
NT2 puerto rico
NT1 north america
NT2 bermuda
NT2 canada
NT2 greenland
NT2 mexico
NT2 st pierre and miquelon
NT2 usa
NT1 south america
NT2 argentina
NT2 bolivia
NT2 brazil
NT2 chile
NT2 colombia

AMERICA (C)
NT2 ecuador
NT2 falkland islands
NT2 french guiana
NT2 guyana
NT2 paraguay
NT2 peru
NT2 surinam
NT2 uruguay
NT2 venezuela

AMORTIZATION
uf depreciation
BT1 book-keeping
NT1 amortization methods

AMORTIZATION METHODS
BT1 amortization
BT2 book-keeping

analysis
SEE **economic analysis**
OR **operations research**
OR **statistical analysis**

ANCILLARY ENTERPRISES
(non farm activities carried out on farm)
BT1 farm enterprises
rt part-time farming
rt tourism

ANDORRA
BT1 western europe
BT2 europe
rt france
rt spain

ANGOLA
BT1 southern africa
BT2 africa south of sahara
BT3 africa
rt portugal

anguilla, nevis
USE **st kitts and nevis and anguilla**

ANIMAL BREEDING
sf breeding
BT1 livestock sector
NT1 artificial insemination
NT1 breeds
NT1 cross breeding
NT1 mating
NT1 selective breeding
rt livestock enterprises

ANIMAL DISEASES
sf diseases
rt parasites
rt public health legislation
rt veterinary services

ANIMAL FATS
(includes animal oils)
BT1 oils and fats

animal husbandry
USE **livestock husbandry**

ANIMAL PROTEINS
BT1 proteins

ANIMAL WELFARE
BT1 livestock sector

animals
SEE **draught animals**
OR **livestock**

ANNUAL ACCOUNTS
BT1 balance sheet analysis

ANTARCTICA
rt australia
rt france
rt norway
rt united kingdom

ANTIBIOTICS
BT1 additives

ANTIGUA
BT1 west indies associated states
BT2 caribbean
BT3 america

antilles
USE **caribbean**

antilles (netherlands)
USE **netherlands antilles**

APICULTURE
uf beekeeping
BT1 livestock enterprises
BT2 farm enterprises
rt bees
rt honey

APPLES
BT1 pome fruits
BT2 fruits

APPLIED RESEARCH
BT1 research

APPRENTICESHIP
BT1 vocational training
BT2 education

APPRENTICESHIP ARTICLES
BT1 educational policy

APPROPRIATE TECHNOLOGY
uf intermediate technology
uf low cost technology
BT1 technical progress
BT2 modernization
BT3 social change

APRICOTS
BT1 stone fruits
BT2 fruits

AQUACULTURE
BT1 livestock enterprises
BT2 farm enterprises
rt shellfish

ARAB COUNTRIES
uf arab states
BT1 algeria
NT1 egypt
NT1 iraq
NT1 jordan
NT1 lebanon
NT1 libya
NT1 morocco
NT1 saudi arabia
NT1 sudan
NT1 syria
NT1 tunisia
NT1 yemen
NT1 yemen pdr
rt north africa

arab republic of egypt
USE **egypt**

arab states
USE **arab countries**

arable farming
USE **crop enterprises**

ARABLE LAND
BT1 farm area
BT1 land

ARGENTINA
BT1 south america
BT2 america
rt lafta

ARID ZONES
BT1 climatic areas

ARID ZONES (C)
NT1 deserts
NT2 desertification

ARTICHOKES
BT1 vegetables

ARTIFICIAL INSEMINATION
BT1 animal breeding
BT2 livestock sector

ARTIFICIAL SWEETENERS
uf sweeteners
BT1 simulated foods
rt isomerose
rt sugar

ASIA
NT1 east asia
NT2 china
NT2 hong kong
NT2 japan
NT2 korea dpr
NT2 korea r
NT2 macao
NT2 mongolia
NT1 persian gulf states
NT2 bahrain
NT2 kuwait
NT2 qatar
NT2 united arab emirates
NT1 south east asia
NT2 bangladesh
NT2 bhutan
NT2 brunei
NT2 burma
NT2 india
NT2 indochina
NT3 kampuchea
NT3 lao
NT3 vietnam
NT2 indonesia
NT2 malaysia
NT2 maldives
NT2 nepal
NT2 pakistan
NT2 philippines
NT2 portuguese timor
NT2 singapore
NT2 sri lanka
NT2 thailand
NT1 west asia
NT2 afghanistan
NT2 cyprus
NT2 iran
NT2 iraq
NT2 israel
NT2 jordan
NT2 lebanon
NT2 oman
NT2 saudi arabia
NT2 syria
NT2 yemen
NT2 yemen pdr

ASPARAGUS
BT1 vegetables

ASPIRATIONS
BT1 roles
BT2 social stratification
BT3 social structure

ASSETS
BT1 balance sheets
BT2 balance sheet analysis
NT1 fixed capital
NT1 livestock capital
NT1 working capital
NT2 liquid assets

associations
SEE **farmers associations**
OR **organizations**
OR **trade associations**

atlases
USE **maps**

ATTITUDES
BT1 behaviour
NT1 political attitudes

ATTITUDES TO WORK
BT1 working conditions

AUBERGINES
BT1 vegetables

AUCTIONS
BT1 marketing techniques

AUDIOVISUAL AIDS
BT1 teaching methods
NT1 educational radio
NT1 educational television
rt films
rt radio

australasia
USE **oceania**

AUSTRALIA
BT1 oceania
rt antarctica
rt commonwealth
rt pacific islands au

AUSTRIA
BT1 western europe
BT2 europe
rt efta

AUTHORITY
BT1 power
BT2 social systems

AUTOMATION
rt computers
rt cybernetics
rt mechanization
rt technical progress

AVERAGE PRICES
BT1 prices

AVERAGES
BT1 central value characteristics
BT2 descriptive statistics
BT3 statistics

aviation
SEE **agricultural aviation**
OR **air transport**

AVOCADOES
BT1 tropical fruits
BT2 fruits

AZORES
BT1 western europe
BT2 europe
rt portugal

baby foods
USE **infant foods**

BACON
uf *smoked bacon*
BT1 meat products
rt pigmeat

BACTERICIDES
BT1 pesticides

BAHAMAS
BT1 caribbean
BT2 america
rt commonwealth
rt united kingdom

BAHRAIN
BT1 persian gulf states
BT2 asia

BAKERY INDUSTRY
BT1 food industry

BAKERY INDUSTRY (C)
NT1 pasta industry
rt bakery products

BAKERY PRODUCTS
NT1 bread
rt bakery industry
rt cereal products
rt food

BALANCE ADJUSTMENT ITEMS
BT1 liabilities
BT2 balance sheets
BT3 balance sheet analysis
NT1 consolidated balance sheets

BALANCE OF PAYMENTS
NT1 terms of trade
rt balance of trade
rt external debt

BALANCE OF TRADE
BT1 trade policy
rt balance of payments
rt supply balance

BALANCE SHEET ANALYSIS
NT1 accounting balance
NT1 annual accounts
NT1 balance sheets
NT2 assets
NT3 fixed capital
NT3 livestock capital
NT3 working capital
NT4 liquid assets
NT2 cash flow
NT2 liabilities
NT3 balance adjustment items
NT4 consolidated balance sheets
NT3 outside capital
NT3 owners equity
NT3 reserve funds
NT1 replacement value
rt accounting

BALANCE SHEETS
BT1 balance sheet analysis
NT1 assets
NT2 fixed capital
NT2 livestock capital
NT2 working capital
NT3 liquid assets
NT1 cash flow
NT1 liabilities
NT2 balance adjustment items
NT3 consolidated balance sheets
NT2 outside capital
NT2 owners equity
NT2 reserve funds

bales
USE **packs**

BANANAS
BT1 tropical fruits
BT2 fruits

BANGLADESH
BT1 south east asia
BT2 asia
rt commonwealth

BANK LOANS
BT1 private loans
BT2 loans
BT3 credit transactions

BANKRUPTCY
BT1 farm indebtedness
BT2 agricultural situation
BT2 farm results

BANKS
NT1 agricultural banks
NT1 co-operative banks
NT2 co-operative credit
NT1 commercial banks
NT1 development banks

BANKS (C)
NT1 industrial banks
NT1 investment banks
NT1 savings banks
rt financial institutions

BARBADOS
BT1 caribbean
BT2 america
rt commonwealth

BARLEY
BT1 cereals
BT2 crops
NT1 barley for feed
NT1 malting barley

BARLEY FOR FEED
BT1 barley
BT2 cereals
BT3 crops

BARTER
BT1 trade

basic funds
USE **fixed capital**

BASIC NEEDS
BT1 living standards
BT2 living conditions

BASIC PRICES
BT1 prices

BASIC RESEARCH
BT1 research

BAYESIAN THEORY
sf *theory*
BT1 operations research
rt decision making

BEANS
BT1 legumes
BT2 vegetables
NT1 broad beans
NT1 french beans
NT1 horse beans
NT1 kidney beans
NT1 lima beans
NT1 white beans

BEEF
BT1 meat

BEEF CATTLE
uf *feeder cattle*
BT1 cattle

beekeeping
USE **apiculture**

BEER
BT1 alcoholic beverages
BT2 beverages
rt brewing industry

BEES
rt apiculture
rt honey

BEET
BT1 vegetables

BEET SUGAR
BT1 sugar
rt sugar beet

BEHAVIOUR
uf *behaviour theory*
NT1 attitudes
NT2 political attitudes
NT1 collective behaviour
NT1 consciousness
NT1 ideology
NT1 opinions
NT1 stereotypes

BEHAVIOUR (C)
rt consumer behaviour
rt farmers attitudes

behaviour theory
USE **behaviour**

BELGIUM
sf benelux
BT1 western europe
BT2 europe
NT1 belgium (antwerp)
NT1 belgium (brabant)
NT1 belgium (east flanders)
NT1 belgium (hainault)
NT1 belgium (liege)
NT1 belgium (limbourg)
NT1 belgium (namur)
NT1 belgium (west flanders)
rt european communities
rt luxembourg
rt netherlands

BELGIUM (ANTWERP)
BT1 belgium
BT2 western europe
BT3 europe

BELGIUM (BRABANT)
BT1 belgium
BT2 western europe
BT3 europe

BELGIUM (EAST FLANDERS)
BT1 belgium
BT2 western europe
BT3 europe

BELGIUM (HAINAULT)
BT1 belgium
BT2 western europe
BT3 europe

BELGIUM (LIEGE)
BT1 belgium
BT2 western europe
BT3 europe

BELGIUM (LIMBOURG)
BT1 belgium
BT2 western europe
BT3 europe

BELGIUM (NAMUR)
BT1 belgium
BT2 western europe
BT3 europe

BELGIUM (WEST FLANDERS)
BT1 belgium
BT2 western europe
BT3 europe

BELIZE
uf british honduras
uf honduras (british)
BT1 central america
BT2 america
rt united kingdom

benelux
SEE **belgium**
OR **luxembourg**
OR **netherlands**

BENIN
uf dahomey
BT1 west africa
BT2 africa south of sahara
BT3 africa
rt french-speaking africa

BERMUDA
BT1 north america
BT2 america
rt united kingdom

BERRIES
BT1 fruits
NT1 bilberries
NT1 blackberries
NT1 blackcurrants
NT1 gooseberries
NT1 raspberries
NT1 redcurrants
NT1 strawberries

BEVERAGE INDUSTRY
BT1 food industry
NT1 brewing industry
NT1 coffee industry
NT1 distilling industry
NT1 tea industry
NT1 wine industry

BEVERAGES
uf chocolate beverages+
uf cocoa beverages+
uf drinking chocolate+
NT1 alcoholic beverages
NT2 beer
NT2 cider
NT2 liqueurs
NT2 spirits
NT2 wine
NT3 champagne and sparkling wines
NT3 quality wine
NT3 table wine
NT1 soft drinks
NT2 coffee
NT2 mate
NT2 tea
rt food

BHUTAN
BT1 south east asia
BT2 asia

BIBLIOGRAPHIES
BT1 secondary journals
BT2 periodicals
BT3 publications

BILBERRIES
uf blueberries
BT1 berries
BT2 fruits

BILLY GOATS
BT1 goats

BIOLOGICAL CONTROL
BT1 plant husbandry
BT2 cultivation methods

birth control
USE **family planning**

BIRTH RATE
BT1 population change
rt sexual behaviour

BITTER ALMONDS
BT1 almonds
BT2 nuts

black tea
USE **tea**

BLACKBERRIES
BT1 berries
BT2 fruits

BLACKCURRANTS
BT1 berries
BT2 fruits

BLOCKED CURRENCIES
uf non-convertible currencies
BT1 currency

blueberries
USE **bilberries**

BOARS
BT1 pigs

BOLIVIA
BT1 south america
BT2 america
rt lafta

BONUSES
BT1 marketing techniques

BOOK-KEEPING
NT1 amortization
NT2 amortization methods
NT1 book-keeping systems
NT2 double entry book-keeping
NT2 global accounting
NT2 single entry book-keeping
NT1 inventories
NT1 plan comptable
NT1 stock accounting
rt accounting

BOOK-KEEPING SYSTEMS
BT1 book-keeping
NT1 double entry book-keeping
NT1 global accounting
NT1 single entry book-keeping

BOOKS
BT1 publications

BOOLEAN ALGEBRA
BT1 mathematics
rt electronic data processing

borneo
USE **indonesia**

BOTSWANA
BT1 southern africa
BT2 africa south of sahara
BT3 africa
rt commonwealth
rt english-speaking africa

BOTTLING
BT1 packaging

BOYCOTTS
BT1 discrimination
BT2 protection

brands
SEE **labelling**
OR **trade marks**

BRAZIL
BT1 south america
BT2 america
rt lafta

BRAZIL NUTS
BT1 nuts

brd
USE **germany fr**

BREAD
BT1 bakery products

BREAK-EVEN POINT
BT1 profitability
BT2 profit and loss analysis

BREEDERS' IMPROVEMENT SOCIETIES
BT1 co-operative services
rt breeds

breeding
SEE **animal breeding**
OR **plant breeding**

BREEDS
BT1 animal breeding
BT2 livestock sector
rt breeders' improvement societies

BREWING INDUSTRY
BT1 beverage industry
BT2 food industry
rt beer

british honduras
USE **belize**

british virgin islands
USE **tortola**

BROAD BEANS
BT1 beans
BT2 legumes
BT3 vegetables

BROCCOLI
BT1 vegetables

BROILERS
uf *chickens+*
BT1 poultry
NT1 cockerels
NT1 day old chicks
NT1 hens

BRUNEI
BT1 south east asia
BT2 asia
rt united kingdom

BRUSSELS SPROUTS
BT1 vegetables

BUCKWHEAT
BT1 cereals
BT2 crops

BUDGET ACCOUNTING
sf *budgeting*
BT1 financial planning
BT2 farm planning

BUDGETARY CONTROL
BT1 project control
BT2 operations research

budgeting
SEE **budget accounting**
OR **farm budgeting**

BUFFALOES
uf *water buffaloes*
BT1 cattle

BUFFER STOCKS
uf *reserve stocks*
BT1 stocks
rt market stabilization

BUILDING CONTROLS
BT1 rural planning
BT2 planning
rt building permits
rt livestock housing
rt zoning

BUILDING PERMITS
BT1 licenses and permits
rt building controls

buildings
SEE **farm buildings**
OR **farm dwellings**
OR **livestock housing**
OR **rural housing**

BULBS
BT1 flowers
BT2 ornamental plants
BT3 crops

BULGARIA
BT1 eastern europe
BT2 europe
rt cmea

BULK HANDLING
BT1 handling

BULK STORAGE
BT1 storage

BULLOCKS
uf *steers*
BT1 cattle

BULLS
BT1 cattle

BURMA
BT1 south east asia
BT2 asia

BURNING
BT1 plant protection

BURUNDI
BT1 central africa
BT2 africa south of sahara
BT3 africa
BT1 east africa
rt french-speaking africa

BUTTER
uf *farm butter*
BT1 milk products
rt margarine

butterfat
USE **milk fat**

BUTTERMILK
(includes buttermilk originating from buttermaking process)
BT1 cultured milks
BT2 milk products

BUYING GROUPS
BT1 wholesale marketing
BT2 marketing channels
BT3 marketing
rt co-operative purchasing

BUYING PRICES
BT1 prices

BY-PRODUCTS
BT1 products

BYELORUSSIAN SSR
BT1 eastern europe
BT2 europe
rt ussr

CABBAGES
BT1 vegetables
NT1 green cabbages
NT1 red cabbages
NT1 savoy cabbages
NT1 white cabbages

cacao
USE **cocoa**

CACM
(central american common market)
BT1 organizations
rt costa rica
rt el salvador
rt guatemala
rt honduras
rt nicaragua

CADASTERS
uf *land registers*
rt land ownership

caicos
USE **turks and caicos islands**

CALCULUS
BT1 mathematics

CALVES
BT1 cattle

cambodia
USE **kampuchea**

CAMEROON
BT1 central africa
BT2 africa south of sahara
BT3 africa
BT1 west africa
rt french-speaking africa

CANADA
BT1 north america
BT2 america
rt commonwealth

CANARY ISLANDS
BT1 western europe
BT2 europe
rt spain

CANE SUGAR
BT1 sugar
rt sugar cane

canned meals
USE **prepared foods**

CANNING
BT1 packaging

CANNING INDUSTRY
BT1 food industry

cantaloupes
USE **melons**

CAP
uf *common agricultural policy*
BT1 agricultural policy
rt european communities
rt less favoured areas
rt policy

CAPE VERDE
BT1 west africa
BT2 africa south of sahara
BT3 africa
rt portugal

CAPITAL
BT1 economic resources
BT2 economic situation
rt factors of production
rt fixed capital
rt investment
rt working capital

CAPITAL ACCOUNTS
BT1 economic accounts

CAPITAL ALLOCATION
BT1 agricultural situation

capital gains tax
USE **capital taxation**

CAPITAL LEASING
sf *leasing*
BT1 commercial credit
BT2 credit
BT3 credit transactions
BT2 trade policy

CAPITAL MARKET
BT1 markets

CAPITAL PRODUCTIVITY
BT1 productivity
BT2 agricultural situation
BT2 profit and loss analysis

CAPITAL TAXATION
uf *capital gains tax*
uf *capital transfer tax*
uf *wealth tax*

CAPITAL TAXATION (C)
sf *taxation*
BT1 direct taxation
BT2 fiscal policy
BT3 public finance
NT1 land tax

capital transfer tax
USE **capital taxation**

CAPITALISM
BT1 political systems

capsicums
USE **peppers**

CARD SERVICES
BT1 information retrieval

CARIBBEAN
uf *antilles*
uf *west indies*
BT1 america
NT1 bahamas
NT1 barbados
NT1 cayman islands
NT1 cuba
NT1 dominican republic
NT1 guadeloupe
NT1 haiti
NT1 jamaica
NT1 martinique
NT1 montserrat
NT1 netherlands antilles
NT1 puerto rico
NT1 tortola
NT1 trinidad and tobago
NT1 turks and caicos islands
NT1 united states virgin islands
NT1 west indies associated states
NT2 antigua
NT2 dominica
NT2 grenada
NT2 st kitts and nevis and anguilla
NT2 st lucia
NT2 st vincent
rt central america
rt south america

CARROTS
BT1 vegetables

cartons
USE **packs**

CASE LAW
NT1 jurisprudence
NT1 legal principles
NT1 practice

CASE STUDIES
BT1 teaching methods

CASH FLOW
BT1 balance sheets
BT2 balance sheet analysis

CASH FLOW ANALYSIS
BT1 plan implementation and evaluation
BT2 operations research

CASHEW NUTS
BT1 nuts

cassava
USE **manioc**

CASTOR BEANS
BT1 oilseeds
BT2 crops

CATCH CROPS
BT1 crops
rt rotation

catchment areas
USE **watersheds**

CATERING INDUSTRY
BT1 food industry

CATTLE
NT1 beef cattle
NT1 buffaloes
NT1 bullocks
NT1 bulls
NT1 calves
NT1 cows
NT2 cows for beef
NT2 dairy cattle
NT2 nurse cows
NT1 draught animals
NT1 dual purpose cattle
NT1 heifers
rt cattle farming
rt livestock

CATTLE FARMING
BT1 livestock enterprises
BT2 farm enterprises
NT1 dairy farming
rt cattle

CAULIFLOWERS
BT1 vegetables

CAYMAN ISLANDS
uf *grand cayman*
BT1 caribbean
BT2 america
rt united kingdom

CELERIAC
BT1 vegetables

CELERY
BT1 vegetables

CENSUSES
NT1 agricultural censuses
NT1 population censuses
rt data collection
rt statistics
rt surveys

CENTRAL AFRICA
BT1 africa south of sahara
BT2 africa
NT1 burundi
NT1 cameroon
NT1 central african empire
NT1 chad
NT1 congo
NT1 equatorial guinea
NT1 gabon
NT1 rwanda
NT1 sao tome and principe
NT1 zaire

CENTRAL AFRICAN EMPIRE
uf *central african republic*
BT1 central africa
BT2 africa south of sahara
BT3 africa
rt french-speaking africa

central african republic
USE **central african empire**

CENTRAL AMERICA
sf *latin america*
BT1 america
NT1 belize
NT1 costa rica
NT1 cuba
NT1 dominican republic
NT1 el salvador
NT1 guatemala
NT1 haiti
NT1 honduras
NT1 jamaica
NT1 nicaragua
NT1 panama
NT1 panama canal zone
NT1 puerto rico
rt caribbean

CENTRAL AMERICA (C)
rt south america

CENTRAL GOVERNMENT
BT1 administration
NT1 ministries of agriculture

CENTRAL PLACES
BT1 rural planning
BT2 planning

CENTRAL VALUE CHARACTERISTICS
BT1 descriptive statistics
BT2 statistics
NT1 averages

CENTRAL-REGIONAL RELATIONS

CENTRALISATION
BT1 socioeconomic organization

CENTRALLY PLANNED ECONOMIES
BT1 economic systems

CEREAL PRODUCTS
NT1 cornflakes
NT1 flour
NT2 cornflour
NT2 oatmeal
NT2 rice flour
NT2 rye flour
NT2 wheat flour
NT1 pasta
NT1 pearl barley
NT1 porridge oats
rt bakery products
rt cereals
rt food

CEREALS
uf *grain*
BT1 crops
NT1 barley
NT2 barley for feed
NT2 malting barley
NT1 buckwheat
NT1 feed grains
NT2 coarse grains
NT1 food grains
NT1 maize
NT2 grain maize
NT2 silo maize
NT1 millet
NT1 oats
NT1 rice
NT1 rye
NT1 sorghum
NT1 triticale
NT1 wheat
NT2 hard wheat
NT2 soft wheat
rt cereal products

ceylon
USE **sri lanka**

CHAD
uf *tchad*
BT1 central africa
BT2 africa south of sahara
BT3 africa
rt french-speaking africa

CHAMPAGNE AND SPARKLING WINES
BT1 wine
BT2 alcoholic beverages
BT3 beverages

CHANGE
uf *land development+*
rt population change
rt social change
rt trends

CHARACTER
BT1 individuals

CHEESE
BT1 milk products
NT1 hard cheese
NT1 processed cheese
NT1 soft cheese

CHEMICAL INDUSTRY
BT1 input industries
NT1 fertilizer industry

chemical residues
USE **noxious substances**

CHERRIES
BT1 stone fruits
BT2 fruits

CHESTNUTS
BT1 nuts

CHI SQUARE TEST
BT1 statistical inference
BT2 statistical analysis
BT3 statistics

CHICK PEAS
BT1 vegetables

chickens+
USE **broilers**
AND **hens**

CHICORY
BT1 vegetables

CHILD LABOUR
BT1 agricultural manpower

CHILDHOOD
BT1 age

CHILDREN
BT1 family size
BT2 families
NT1 daughters
NT1 sons

CHILE
BT1 south america
BT2 america
rt lafta

CHINA
uf *china mainland*
uf *china pr*
uf *mainland china*
uf *taiwan*
BT1 east asia
BT2 asia

china mainland
USE **china**

china pr
USE **china**

CHOCOLATE
uf *chocolate beverages+*
uf *drinking chocolate+*
rt food
rt sugar confectionery

chocolate beverages+
USE **beverages**
AND **chocolate**

chocolate industry
USE **cocoa industry**

CHURCHES
BT1 religions
BT2 social systems
NT1 churches in rural areas

CHURCHES IN RURAL AREAS
BT1 churches
BT2 religions
BT3 social systems

CIDER
BT1 alcoholic beverages
BT2 beverages

circulating funds
USE **working capital**

CITRUS FRUITS
BT1 fruits
NT1 clementines
NT1 grapefruits
NT1 lemons
NT1 mandarins
NT1 oranges

civil code
USE **civil law**

CIVIL LAW
uf *civil code*
BT1 legislation
BT2 legal sources

CIVILIZATION
BT1 culture

CLASS CONFLICT
BT1 social unrest
BT2 social crisis
BT3 socialization

CLASSICAL ECONOMICS
uf *liberalism*
BT1 economic schools
BT2 economic theories

CLASSIFICATION
BT1 document analysis
BT2 information processing

CLEMENTINES
BT1 citrus fruits
BT2 fruits

CLIMATE
uf *weather*
NT1 microclimate
rt climatic areas
rt meteorology

CLIMATIC AREAS
NT1 arid zones
NT2 deserts
NT3 desertification
NT1 humid zones
NT1 temperate zones
NT1 tropics
rt climate

CLIMATIC FACTORS
NT1 drought
NT1 frost
NT1 precipitation
NT2 hail
NT2 rainfall
NT2 snow
rt agricultural disasters
rt damage

CLOVER
BT1 feed crops
BT2 crops

CLUBS
BT1 groups

CMEA
(council for mutual economic assistance)
uf *comecon*
BT1 organizations
rt bulgaria
rt cuba
rt czechoslovakia
rt germany dr
rt hungary
rt mongolia
rt poland
rt romania

CMEA (C)
rt ussr

CO-OPERATION
sf *mutual help*
BT1 socioeconomic organization
rt co-operative farm enterprises
rt co-operative services
rt co-operatives
rt co-ownership
rt communes
rt integration
rt socialization

CO-OPERATIVE ACTIVITIES
NT1 co-operative bargaining
NT1 co-operative farming
NT1 co-operative marketing
NT2 co-operative purchasing
NT2 co-operative selling
NT1 co-operative processing
rt co-operatives

CO-OPERATIVE BANKS
BT1 banks
BT1 co-operative services
NT1 co-operative credit

CO-OPERATIVE BARGAINING
BT1 co-operative activities

CO-OPERATIVE CREDIT
BT1 co-operative banks
BT2 banks
BT2 co-operative services
rt agricultural credit

CO-OPERATIVE FARM ENTERPRISES
(cooperation on a single crop or livestock activity)
uf *inter co-operative enterprises*
NT1 crop production groups
NT1 forestry groups
NT1 gaec
NT1 grazing groups
NT1 groupement foncier agricole
NT1 livestock groups
NT1 producer groups
NT1 safer
rt co-operation
rt co-operative farming
rt collective farming

CO-OPERATIVE FARMING
BT1 co-operative activities
rt co-operative farm enterprises
rt collective farms

CO-OPERATIVE FEEDSTORES
(interfarm sale of farmgrown feed on contract)
BT1 farm helper services
BT2 co-operative services

CO-OPERATIVE INSURANCE
BT1 co-operative services

CO-OPERATIVE MARKETING
uf *marketing co-operatives*
BT1 co-operative activities
NT1 co-operative purchasing
NT1 co-operative selling

CO-OPERATIVE PLANT PROTECTION
BT1 co-operative services

CO-OPERATIVE PROCESSING
BT1 co-operative activities

CO-OPERATIVE PURCHASING
uf *supply co-operatives*
BT1 co-operative marketing
BT2 co-operative activities
rt buying groups

CO-OPERATIVE SELLING
BT1 co-operative marketing
BT2 co-operative activities

CO-OPERATIVE SERVICES
NT1 breeders' improvement societies
NT1 co-operative banks
NT2 co-operative credit
NT1 co-operative insurance
NT1 co-operative plant protection
NT1 farm helper services
NT2 co-operative feedstores
NT2 machinery co-operatives
NT2 work banks
NT1 land improvment co-operatives
rt co-operation

CO-OPERATIVES
NT1 consumer co-operatives
NT1 dairy co-operatives
NT1 multi-purpose co-operatives
NT1 mutual organisations
NT1 producer co-operatives
NT1 viticulture co-operatives
rt co-operation
rt co-operative activities
rt private firms

CO-OWNERSHIP
uf *communal ownership*
BT1 ownership
rt co-operation

COARSE GRAINS
BT1 feed grains
BT2 cereals
BT3 crops

COCKERELS
BT1 broilers
BT2 poultry

COCOA
uf *cacao*
uf *cocoa beverages+*
rt food
rt instant cocoa

cocoa beverages+
USE **beverages**
AND **cocoa**

COCOA INDUSTRY
uf *chocolate industry*
BT1 food industry

COCONUT OILS
uf *lauric oils*
BT1 vegetable oils
BT2 oils and fats

COCONUTS
BT1 nuts

CODE FORESTIER
(in france only)
BT1 legislation
BT2 legal sources

CODIFIED RURAL LEGISLATION
BT1 legislation
BT2 legal sources

COFFEE
BT1 soft drinks
BT2 beverages
rt instant coffee

COFFEE INDUSTRY
BT1 beverage industry
BT2 food industry

coffee powder
USE **instant coffee**

coffee whiteners
USE **milk substitutes**

COGNITION
BT1 decision making

COLA NUTS
BT1 nuts

COLD STORAGE
BT1 storage

COLLECTION
BT1 transport

COLLECTIVE AGREEMENTS
BT1 legal sources

COLLECTIVE BEHAVIOUR
BT1 behaviour
rt political attitudes

COLLECTIVE FARMING
BT1 types of farming
BT2 agricultural systems
rt co-operative farm enterprises

COLLECTIVE FARMS
BT1 farms
NT1 kibbutzim
NT1 kolkhozy
NT1 moshavim
rt co-operative farming
rt collectivization
rt communes
rt producer co-operatives

COLLECTIVIZATION
BT1 socialization
rt collective farms

COLOMBIA
BT1 south america
BT2 america
rt lafta

COLONIALISM
BT1 imperialism
BT2 political systems

COLOURINGS
BT1 additives

COMBINE HARVESTER
BT1 harvester
BT2 farm machinery
BT3 farm equipment

comecon
USE **cmea**

COMISSION AGENTS
BT1 merchants
BT2 marketing channels
BT3 marketing

COMMERCIAL BANKS
BT1 banks

COMMERCIAL CREDIT
BT1 credit
BT2 credit transactions
BT1 trade policy
NT1 capital leasing

COMMERCIAL FARMING
BT1 agricultural systems

COMMERCIAL LAW
rt company law
rt legal form
rt trade

COMMODITIES
BT1 products

COMMODITY EXCHANGES
BT1 marketing channels
BT2 marketing
NT1 futures markets

COMMODITY MARKETS
BT1 markets

common agricultural policy
USE **cap**

common grazing
USE **common lands**

common guarantee policy
USE **marketing policy**

common guidance policy
USE **structural policy**

COMMON LANDS
uf *common grazing*
BT1 public domain
BT2 land use systems

common law
USE **customary law**

COMMON MARKETS
BT1 economic unions

COMMONWEALTH
rt australia
rt bahamas
rt bangladesh
rt barbados
rt botswana
rt canada
rt cyprus
rt fiji
rt gambia
rt ghana
rt guyana
rt india
rt jamaica
rt kenya
rt lesotho
rt malawi
rt malaysia
rt malta
rt mauritius
rt nauru
rt new zealand
rt nigeria
rt papua new guinea
rt pomegranates
rt sierra leone
rt singapore
rt sri lanka
rt swaziland
rt tonga
rt trinidad and tobago
rt uganda
rt united kingdom
rt western samoa
rt zambia

communal ownership
USE **co-ownership**

COMMUNES
uf *ujamaa villages*
rt co-operation
rt collective farms
rt groups

COMMUNICATION
rt extension
rt information diffusion
rt mass media

COMMUNISM
BT1 political systems

COMMUNITY DEVELOPMENT
BT1 rural development
BT2 rural environment
BT3 environment
NT1 rural animation

COMMUTING
BT1 geographical mobility
BT2 migration

COMORO ISLANDS
BT1 southern africa

COMORO ISLANDS (C)
BT2 africa south of sahara
BT3 africa
rt france
rt french-speaking africa

COMPANIES
BT1 private firms
BT2 firms
rt company law
rt legal form
rt mutual organisations

COMPANY LAW
rt commercial law
rt companies
rt legal form

company statutes
USE **legal form**

COMPANY TAX
BT1 direct taxation
BT2 fiscal policy
BT3 public finance

comparative advantage
USE **terms of trade**

COMPENSATION
rt agricultural disasters
rt compulsory purchase
rt insurance benefits
rt valuation

COMPENSATORY AMOUNTS
uf *compensatory monetary amounts*
uf *monetary compensatory amounts*
BT1 price policy
BT2 marketing policy
BT3 agricultural policy

compensatory monetary amounts
USE **compensatory amounts**

COMPETITION
BT1 market economics
NT1 imperfect competition
NT2 monopoly
NT2 monopsony
NT2 oligopoly
NT2 oligopsony
NT1 perfect competition
rt share of the market

COMPETITIVE ECONOMIES
BT1 market economies
BT2 economic systems

COMPONENT ANALYSIS
BT1 statistical inference
BT2 statistical analysis
BT3 statistics
rt factorial analysis

COMPOST
BT1 inorganic fertilizers
BT2 fertilizers

COMPOUND FERTILIZERS
BT1 inorganic fertilizers
BT2 fertilizers

COMPULSORY PURCHASE
BT1 ownership limitation
BT2 ownership
rt compensation
rt land transfers

COMPUTER PROGRAMMING
BT1 programming

COMPUTER SOFTWARE
uf *computer software package*
rt computers
rt electronic data processing

computer software package
USE **computer software**

COMPUTERS
BT1 hardware
NT1 minicomputers
rt automation
rt computer software

CONCAVE PROGRAMMING
BT1 programming

CONCENTRATES
uf *protein feeds*
BT1 manufactured feeds
BT2 feedingstuffs

CONCENTRATION
(concentration of production)
BT1 socioeconomic organization
NT1 farm amalgamation

condensed milk
SEE **evaporated milk**
OR **sweetened condensed milk**

CONFECTIONERY INDUSTRY
BT1 food industry

CONFLICT
BT1 social crisis
BT2 socialization
rt generation conflict

CONGLOMERATES
BT1 firms

CONGO
uf *congo (brazzaville)*
uf *congo pr*
BT1 central africa
BT2 africa south of sahara
BT3 africa
rt french-speaking africa

congo (brazzaville)
USE **congo**

congo (kinshasa)
USE **zaire**

congo (leopoldville)
USE **zaire**

congo democratic republic
USE **zaire**

congo pr
USE **congo**

CONGRESSES
rt publications
rt trade associations

CONSCIOUSNESS
BT1 behaviour

CONSOLIDATED BALANCE SHEETS
BT1 balance adjustment items
BT2 liabilities
BT3 balance sheets
BT4 balance sheet analysis

CONSTITUTIONAL LAW
BT1 legal sources

CONSUMER BEHAVIOUR
uf *sensory evaluation+*
BT1 economic sociology
NT1 eating habits
NT1 purchasing habits
rt behaviour
rt consumer surveys
rt household consumption

CONSUMER CO-OPERATIVES
BT1 co-operatives

CONSUMER EXPENDITURE
BT1 aggregate data
NT1 household expenditure
rt consumer surveys
rt economic sociology
rt household consumption

CONSUMER INFORMATION
BT1 consumer protection

CONSUMER MOTIVATION
BT1 consumer surveys
BT2 market research
rt economic sociology

CONSUMER PANELS
BT1 consumer surveys
BT2 market research

CONSUMER PRICES
BT1 prices
rt food prices
rt retail prices

CONSUMER PROTECTION
NT1 consumer information
NT1 prevention of fraud
NT1 quality controls
NT2 designation of origin
NT2 labelling controls
rt food inspection

CONSUMER SURVEYS
BT1 market research
NT1 consumer motivation
NT1 consumer panels
rt consumer behaviour
rt consumer expenditure

CONSUMPTION
NT1 consumption per caput
NT1 consumption units
NT1 domestic consumption
NT1 household consumption
NT1 on farm consumption
rt demand

CONSUMPTION FUNCTIONS
BT1 single equation analysis
BT2 parametric methods
BT3 econometrics

CONSUMPTION PER CAPUT
BT1 consumption

CONSUMPTION UNITS
BT1 consumption

CONTAINER TRANSPORT
BT1 transport

CONTINUING EDUCATION
uf *further education*
BT1 education
rt retraining

CONTRACT FARMING
rt vertical integration

CONTRACT LEGISLATION
rt agricultural law
rt integration
rt social legislation

CONTRACTS
BT1 acquisition of ownership
rt option to purchase

CONTROLLED ATMOSPHERE STORAGE
BT1 storage

CONTROLLED GRAZING
uf *paddock grazing*
BT1 livestock husbandry
BT2 livestock sector
rt fattening
rt grazing

CONTROLLED PRICES
BT1 prices

convenience foods
USE **prepared foods**

CONVEX PROGRAMMING
BT1 programming

COOLING
BT1 preservation

COORDINATION
BT1 socioeconomic organization

CORN OILS
BT1 vegetable oils
BT2 oils and fats

CORNFLAKES
BT1 cereal products

CORNFLOUR
BT1 flour
BT2 cereal products

CORRELATION ANALYSIS
BT1 statistical inference
BT2 statistical analysis
BT3 statistics

CORVICIDES
BT1 pesticides

cost accounting
USE **cost analysis**

COST ANALYSIS
uf *cost accounting*
NT1 costing
NT1 costs
NT2 production costs
NT3 actual costs
NT3 estimated costs
NT3 fixed costs
NT3 special costs
NT3 total costs
NT3 variable costs
NT1 returns
rt accounting

COST BENEFIT ANALYSIS
BT1 non-optimizing methods
BT2 operations research

cost of living
USE **living standards**

COST PRICES
BT1 prices
rt production costs

COSTA RICA
BT1 central america
BT2 america
rt cacm

COSTING
BT1 cost analysis

COSTS
BT1 cost analysis
NT1 production costs
NT2 actual costs
NT2 estimated costs
NT2 fixed costs
NT2 special costs
NT2 total costs
NT2 variable costs

COTTON
BT1 fibre crops
BT2 crops

COTTON GINNING
BT1 cotton industry
BT2 textile industries
BT3 non-food industries

COTTON INDUSTRY
BT1 textile industries
BT2 non-food industries
NT1 cotton ginning

COTTONSEED OILS
BT1 vegetable oils
BT2 oils and fats

COWS
BT1 cattle
NT1 cows for beef
NT1 dairy cattle
NT1 nurse cows

COWS FOR BEEF
BT1 cows
BT2 cattle

cows milk
USE **milk**

CREAM
BT1 milk products

cream substitutes
USE **milk substitutes**

CREDIT
BT1 credit transactions
NT1 agricultural credit
NT1 commercial credit
NT2 capital leasing
NT1 export credits
NT1 long term credit
NT1 medium term credit
NT1 short term credit
rt agricultural debentures
rt loans
rt outside finance

CREDIT CONTROL
BT1 credit policy
NT1 discount rates
NT1 interest rates
NT2 preferential interest rates

credit guaranties
USE **credit insurance**

CREDIT INSURANCE
uf *credit guaranties*
BT1 insurances

CREDIT POLICY
NT1 credit control
NT2 discount rates
NT2 interest rates
NT3 preferential interest rates
NT1 repayment
NT2 debt consolidation
NT2 interest repayments
rt credit transactions
rt loans
rt policy

CREDIT SALES
BT1 marketing techniques

CREDIT TRANSACTIONS
NT1 credit
NT2 agricultural credit
NT2 commercial credit
NT3 capital leasing
NT2 export credits
NT2 long term credit
NT2 medium term credit
NT2 short term credit
NT1 debt
NT2 agricultural debentures
NT2 external debt
NT2 public debt
NT1 hire purchase
NT1 loans
NT2 international loans
NT2 private loans
NT3 bank loans
NT3 mortgages

CREDIT TRANSACTIONS (C)
NT2 public loans
rt agricultural bills of exchange
rt credit policy

CRIMINALITY
BT1 deviant behaviour
BT2 social crisis
BT3 socialization

crisis
USE **economic depression**

CRITICAL PATH ANALYSIS
BT1 plan implementation and evaluation
BT2 operations research

CROP ENTERPRISES
uf *arable farming*
BT1 farm enterprises
NT1 field crops
NT1 forestry
NT1 grazing
NT1 horticulture
NT2 floriculture
NT2 fruit growing
NT2 tree nurseries
NT2 vegetable growing
NT1 ley farming
NT1 viticulture

CROP INSURANCE
uf *storm insurance*
BT1 agricultural insurance
BT2 insurances
rt agricultural disasters

CROP PRODUCTION GROUPS
BT1 co-operative farm enterprises

CROP RESIDUES
BT1 agricultural wastes
BT2 pollution by agriculture
BT3 environmental policy
rt straw

CROPPING SYSTEMS
NT1 dry farming
NT1 greenhouse crops
NT1 hydroponics
NT1 irrigated farming
NT1 organic farming
NT1 perennial cropping
NT1 primitive rotation farming
NT1 terracing
rt crops

CROPS
NT1 catch crops
NT1 cereals
NT2 barley
NT3 barley for feed
NT3 malting barley
NT2 buckwheat
NT2 feed grains
NT3 coarse grains
NT2 food grains
NT2 maize
NT3 grain maize
NT3 silo maize
NT2 millet
NT2 oats
NT2 rice
NT2 rye
NT2 sorghum
NT2 triticale
NT2 wheat
NT3 hard wheat
NT3 soft wheat
NT1 culinary herbs
NT1 feed crops
NT2 clover
NT2 lucerne
NT1 fibre crops
NT2 cotton
NT2 fibre flax
NT2 hemp
NT2 kapok

CROPS (C)
NT2 sisal
NT1 grass
NT1 hops
NT1 medicinal herbs
NT1 oilseeds
NT2 castor beans
NT2 groundnuts
NT2 linseed
NT2 mustard
NT2 rape
NT2 sesame
NT2 soybeans
NT1 ornamental plants
NT2 flowers
NT3 bulbs
NT1 root crops
NT2 fodder beet
NT2 manioc
NT2 potatoes
NT3 feed potatoes
NT3 industrial potatoes
NT3 ware potatoes
NT2 sugar beet
NT2 turnips
NT2 yams
NT1 spice plants
NT1 sugar cane
NT1 trees and shrubs
rt cropping systems
rt field crops

CROSS BREEDING
uf *hybrids*
BT1 animal breeding
BT2 livestock sector
rt hybridization

CROSS SECTIONAL ANALYSIS
BT1 statistical inference
BT2 statistical analysis
BT3 statistics

CUBA
BT1 caribbean
BT2 america
BT1 central america
rt cmea

CUCUMBERS
BT1 vegetables

CULINARY HERBS
sf *herbs*
BT1 crops

cultivars
USE **varieties**

CULTIVATION METHODS
NT1 harvesting
NT1 land clearance
NT1 land improvement
NT2 drainage
NT2 fertilizer application
NT2 irrigation
NT2 soil conservation
NT2 spray irrigation
NT1 plant husbandry
NT2 biological control
NT2 grafting
NT2 integrated plant protection
NT2 pest control
NT2 pruning
NT2 thinning
NT2 weed control
NT1 rotation
NT1 tillage
NT2 harrowing
NT2 minimal tillage systems
NT2 no-tillage system
NT2 planting
NT2 ploughing
NT2 replanting
NT2 rolling
NT2 sowing
NT2 subsoiling
rt land use

CULTURAL DEVELOPMENT
BT1 culture

CULTURAL DIFFERENTIATION
BT1 cultural environment
BT2 culture

CULTURAL ENVIRONMENT
BT1 culture
NT1 acculturation
NT1 cultural differentiation
NT1 cultural integration

CULTURAL HERITAGE
BT1 culture

CULTURAL INTEGRATION
BT1 cultural environment
BT2 culture

CULTURE
NT1 civilization
NT1 cultural development
NT1 cultural environment
NT2 acculturation
NT2 cultural differentiation
NT2 cultural integration
NT1 cultural heritage
NT1 value systems
NT1 values
rt social anthropology

CULTURED MILKS
uf *fermented milk*
BT1 milk products
NT1 buttermilk
NT1 yoghurt

CURRENCY
NT1 blocked currencies
NT1 european units of account
NT1 foreign exchange
NT1 green money
NT1 national currencies

CURRICULUM
rt education
rt literacy

CUSTOMARY LAW
uf *common law*
BT1 legal sources

customs duties
USE **tariffs**

CUSTOMS UNIONS
BT1 economic unions

CYBERNETICS
rt automation
rt mathematics
rt operations research

CYCLIC FLUCTUATIONS
BT1 fluctuations
BT2 trade cycles
BT3 economic development

CYPRUS
BT1 mediterranean countries
BT1 west asia
BT2 asia
BT1 western europe
BT2 europe
rt commonwealth

CZECHOSLOVAKIA
BT1 eastern europe
BT2 europe
rt cmea

dahomey
USE **benin**

DAIRY CATTLE
BT1 cows
BT2 cattle

DAIRY CATTLE (C)
rt dairy farming

DAIRY CO-OPERATIVES
BT1 co-operatives

DAIRY FARMING
BT1 cattle farming
BT2 livestock enterprises
BT3 farm enterprises
rt dairy cattle

DAIRY INDUSTRY
BT1 food industry

DAIRY-BEEF CONVERSION SCHEMES
BT1 adjustment of production
BT2 production policy

DAMAGE
rt agricultural disasters
rt climatic factors
rt subsidies

DATA BANKS
BT1 documentation
rt information storage

DATA BASES
BT1 documentation
rt information storage

DATA COLLECTION
rt censuses
rt data processing
rt sampling
rt surveys

DATA PROCESSING
uf *teleprocessing*
NT1 on line
rt data collection

DATE STAMPING
BT1 standards

DATES
BT1 tropical fruits
BT2 fruits

DAUGHTERS
BT1 children
BT2 family size
BT3 families

DAY OLD CHICKS
BT1 broilers
BT2 poultry

ddr
USE **germany dr**

DEBT
uf *indebtedness*
BT1 credit transactions
NT1 agricultural debentures
NT1 external debt
NT1 public debt
rt farm indebtedness
rt losses
rt outside finance

DEBT CONSOLIDATION
BT1 repayment
BT2 credit policy

debt servicing
SEE **interest repayments**
OR **repayment**

decaffeinated tea
USE **tea**

DECENTRALIZATION
BT1 socioeconomic organization

DECISION MAKING
NT1 action

DECISION MAKING (C)
NT1 cognition
NT1 motivation
NT1 thinking
rt bayesian theory

DECREES
BT1 regulations
BT2 legislation
BT3 legal sources

DEER
BT1 game

DEFLATION
BT1 monetary situation

DEMAND
BT1 market economics
rt consumption
rt household consumption
rt supply and demand
rt supply balance

DEMAND FUNCTIONS
BT1 single equation analysis
BT2 parametric methods
BT3 econometrics

DEMOCRACY
BT1 political systems
NT1 parliament

DEMOGRAPHY
rt population

DENMARK
BT1 scandinavia
BT2 western europe
BT3 europe
NT1 denmark (bornholm)
NT1 denmark (fyn)
NT1 denmark (jutland)
NT1 denmark (lolland-falster)
NT1 denmark (zealand)
rt european communities
rt faroe islands
rt greenland

DENMARK (BORNHOLM)
BT1 denmark
BT2 scandinavia
BT3 western europe
BT4 europe

DENMARK (FYN)
BT1 denmark
BT2 scandinavia
BT3 western europe
BT4 europe

DENMARK (JUTLAND)
BT1 denmark
BT2 scandinavia
BT3 western europe
BT4 europe

DENMARK (LOLLAND-FALSTER)
BT1 denmark
BT2 scandinavia
BT3 western europe
BT4 europe

DENMARK (ZEALAND)
BT1 denmark
BT2 scandinavia
BT3 western europe
BT4 europe

depreciation
USE **amortization**

DESCRIPTIVE STATISTICS
BT1 statistics
NT1 central value characteristics
NT2 averages
NT1 dispersion and distribution
NT2 standard deviation

DESCRIPTIVE STATISTICS (C)
NT2 time series
NT3 seasonality
NT2 variance

DESERTIFICATION
BT1 deserts
BT2 arid zones
BT3 climatic areas

DESERTS
BT1 arid zones
BT2 climatic areas
NT1 desertification
rt nomads

DESIGNATION OF ORIGIN
BT1 quality controls
BT2 consumer protection

DEVALUATION
BT1 monetary parity
BT2 monetary policy

DEVELOPED COUNTRIES
BT1 economic geography

DEVELOPING COUNTRIES
uf *third world*
uf *underdeveloped countries*
sf *less developed countries*
BT1 economic geography
rt development aid

DEVELOPMENT AID
sf *aid*
NT1 food aid
NT1 technical aid
rt developing countries
rt development policy
rt development projects
rt funds
rt international agreements

DEVELOPMENT BANKS
BT1 banks
rt industrial banks
rt investment banks

DEVELOPMENT PLANS
BT1 planning
NT1 development projects

DEVELOPMENT POLICY
rt development aid
rt policy

DEVELOPMENT PROJECTS
BT1 development plans
BT2 planning
rt development aid

DEVELOPMENT THEORY
sf *theory*
BT1 economic theories
NT1 economic dualism

DEVIANT BEHAVIOUR
BT1 social crisis
BT2 socialization
NT1 alcoholism
NT1 criminality

DICTATORSHIP
BT1 political systems

DICTIONARY
BT1 publications

DIETETIC FOODS
BT1 prepared foods
NT1 infant foods

DIFFUSION OF RESEARCH
BT1 research policy
rt documentation
rt extension
rt information diffusion

DIRECT MARKETING
BT1 retail marketing
BT2 marketing channels
BT3 marketing

DIRECT TAXATION
sf *taxation*
BT1 fiscal policy
BT2 public finance
NT1 capital taxation
NT2 land tax
NT1 company tax
NT1 income tax
NT1 inheritance tax

DIRECTORIES
BT1 publications

DISCOUNT RATES
BT1 credit control
BT2 credit policy

DISCOUNTS
BT1 marketing techniques

DISCRIMINANT ANALYSIS
uf *multiple discriminant analysis*
BT1 statistical inference
BT2 statistical analysis
BT3 statistics

DISCRIMINATION
BT1 protection
NT1 boycotts
NT1 dumping
NT1 embargos
NT1 price discrimination

DISCUSSION GROUPS
BT1 teaching methods

diseases
SEE **animal diseases**
OR **plant diseases**

DISPARITY
BT1 agricultural situation
rt income distribution

DISPERSION AND DISTRIBUTION
BT1 descriptive statistics
BT2 statistics
NT1 standard deviation
NT1 time series
NT2 seasonality
NT1 variance

DISTILLING INDUSTRY
BT1 beverage industry
BT2 food industry

DISTRIBUTION
BT1 marketing
rt marketing margins

distribution channels
USE **marketing channels**

DIVERSIFICATION
BT1 socioeconomic organization
rt integration
rt mixed cropping

DIVISION OF LABOUR
BT1 labour economics

DIVORCE
BT1 marriage
BT2 families

DOCUMENT ANALYSIS
BT1 information processing
NT1 abstracting
NT1 classification
NT1 indexing

DOCUMENTATION
NT1 data banks

DOCUMENTATION (C)
NT1 data bases
rt diffusion of research
rt history
rt information diffusion
rt information retrieval
rt rural sociology research

DOCUMENTATION LANGUAGES
BT1 information processing
NT1 thesauri

DOMESTIC CONSUMPTION
BT1 consumption

DOMESTIC GARDENS
uf allotments
uf gardens
uf kitchen gardens
BT1 types of farming
BT2 agricultural systems

DOMESTIC MARKETS
BT1 markets

DOMESTIC PRODUCTION
BT1 production
rt self sufficiency

DOMESTIC TRADE
uf home trade
BT1 trade

DOMINICA
BT1 west indies associated states
BT2 caribbean
BT3 america

DOMINICAN REPUBLIC
BT1 caribbean
BT2 america
BT1 central america

DOOR TO DOOR SALES
BT1 marketing techniques

DOUBLE ENTRY BOOK-KEEPING
BT1 book-keeping systems
BT2 book-keeping

DOWNWARD MOBILITY
BT1 vertical mobility
BT2 social mobility
BT3 social stratification
BT4 social structure

DRAINAGE
BT1 land improvement
BT2 cultivation methods

DRAUGHT ANIMALS
sf animals
BT1 cattle

DRAWING RIGHTS
BT1 monetary systems
rt monetary agreements

DRIED FOODS
BT1 prepared foods
NT1 freeze-dried foods
rt dried fruit
rt instant cocoa
rt instant coffee
rt instant tea

DRIED FRUIT
BT1 fruit products
NT1 prunes
NT1 raisins
rt dried foods
rt fruits

DRIED MEATS
BT1 meat products

DRIED MILK
uf milk powder

DRIED MILK (C)
uf whole milk powder
BT1 milk products
NT1 dried skim milk

DRIED SKIM MILK
BT1 dried milk
BT2 milk products

drinking chocolate+
USE **beverages**
AND **chocolate**

DROUGHT
BT1 climatic factors

DRY FARMING
BT1 cropping systems

DRY FEED
BT1 farm produced fodder
BT2 feedingstuffs

DRYING
BT1 preservation

DUAL PURPOSE CATTLE
BT1 cattle

DUAL PURPOSE SHEEP
BT1 sheep

DUCKS
BT1 poultry

DUMPING
BT1 discrimination
BT2 protection
rt imperfect competition

durum wheat
USE **hard wheat**

DYNAMIC MODELS
BT1 models

DYNAMIC PROGRAMMING
BT1 programming

EAGGF
(european agricultural guidance and guarantee fund)
BT1 funds

EARNED INCOME
BT1 farm results

EASEMENTS
BT1 ownership limitation
BT2 ownership

EAST AFRICA
BT1 africa south of sahara
BT2 africa
NT1 afars and issas territory
NT1 burundi
NT1 ethiopia
NT1 kenya
NT1 malawi
NT1 rwanda
NT1 seychelles
NT1 somalia
NT1 sudan
NT1 tanzania
NT1 uganda

EAST ASIA
uf far east
BT1 asia
NT1 china
NT1 hong kong
NT1 japan
NT1 korea dpr
NT1 korea r
NT1 macao
NT1 mongolia

east berlin
USE **germany dr**

EASTERN EUROPE
BT1 europe
NT1 albania
NT1 bulgaria
NT1 byelorussian ssr
NT1 czechoslovakia
NT1 germany dr
NT1 hungary
NT1 poland
NT1 romania
NT1 ukrainian ssr
NT1 ussr
NT1 yugoslavia

eastern germany
USE **germany dr**

EATING HABITS
BT1 consumer behaviour
BT2 economic sociology
rt nutrition
rt on farm consumption

EC DIRECTIVES

EC LAWS
BT1 legal sources
NT1 treaty of rome
NT2 ec regulations

EC LEGAL DECISIONS
rt practice
rt treaty of rome

EC REGULATIONS
BT1 treaty of rome
BT2 ec laws
BT3 legal sources

ECOLOGY
BT1 environment

ECONOMETRIC MODELS
BT1 models

ECONOMETRICS
NT1 non-parametric methods
NT1 parametric methods
NT2 simultaneous equation analysis
NT2 single equation analysis
NT3 consumption functions
NT3 demand functions
NT3 elasticities
NT3 flexibility
NT3 investment functions
NT3 production functions
NT3 savings functions
NT3 supply functions
NT3 utility functions
rt operations research
rt sociological method

ECONOMIC ACCOUNTS
NT1 capital accounts
NT1 farm accounts
NT1 income accounts
NT1 production accounts
rt aggregate data

ECONOMIC ANALYSIS
sf analysis
NT1 marginal analysis
NT2 marginal costs
NT2 marginal returns
rt agricultural situation
rt gross margins analysis
rt input-output analysis
rt microeconomic analysis

ECONOMIC DEPRESSION
uf crisis
uf recession
BT1 trade cycles
BT2 economic development
rt agricultural crisis

ECONOMIC DEPRESSION (C)
rt social crisis

ECONOMIC DEVELOPMENT
NT1 economic growth
NT2 growth rate
NT1 regional development
NT1 sectoral development
NT1 trade cycles
NT2 economic depression
NT2 economic recovery
NT2 fluctuations
NT3 cyclic fluctuations
NT3 seasonal fluctuations
NT3 trends
NT2 stabilization
NT2 stagnation
rt agrarian reform
rt social change

ECONOMIC DUALISM
BT1 development theory
BT2 economic theories

ECONOMIC GEOGRAPHY
NT1 agrarian countries
NT1 developed countries
NT1 developing countries
NT1 least developed countries
rt geography

ECONOMIC GROWTH
uf growth
BT1 economic development
NT1 growth rate
rt growth theory

ECONOMIC LAWS
BT1 economic theories
NT1 economies of scale

ECONOMIC RECOVERY
BT1 trade cycles
BT2 economic development

ECONOMIC REGIONS
BT1 regions

ECONOMIC RESOURCES
BT1 economic situation
NT1 capital
NT1 energy
NT2 electricity
NT2 solar energy
NT1 human resources
NT2 labour
NT2 management
NT3 farm management
NT3 self management
NT1 natural resources
NT1 production possibilities

ECONOMIC SCHOOLS
BT1 economic theories
NT1 classical economics
NT1 keynsian economics
NT1 marxist economics
NT1 neo-classical economics

ECONOMIC SECTORS
sf sectors of production
NT1 primary sector
NT2 farm sector
NT1 secondary sector
NT1 tertiary sector
NT2 professional services
NT3 veterinary services
NT2 public services
rt sectoral development

ECONOMIC SITUATION
NT1 economic resources
NT2 capital
NT2 energy
NT3 electricity
NT3 solar energy
NT2 human resources
NT3 labour

ECONOMIC SITUATION (C)
NT3 management
NT4 farm management
NT4 self management
NT2 natural resources
NT2 production possibilities

ECONOMIC SOCIOLOGY
NT1 consumer behaviour
NT2 eating habits
NT2 purchasing habits
NT1 farmers attitudes
rt consumer expenditure
rt consumer motivation

ECONOMIC SYSTEMS
NT1 centrally planned economies
NT1 market economies
NT2 competitive economies
NT1 mixed economies
NT2 welfare economics

ECONOMIC THEORIES
sf theory
NT1 development theory
NT2 economic dualism
NT1 economic laws
NT2 economies of scale
NT1 economic schools
NT2 classical economics
NT2 keynsian economics
NT2 marxist economics
NT2 neo-classical economics
NT1 equilibrium theory
NT1 growth theory
NT1 location theory
rt economics
rt market economics

ECONOMIC UNIONS
NT1 common markets
NT1 customs unions
rt monetary unions

ECONOMIC UNITS
NT1 financial institutions
NT1 households
NT2 agricultural households
NT1 production unit
rt manpower units

ECONOMICS
NT1 agricultural economics
rt economic theories

ECONOMIES OF SCALE
uf scale
BT1 economic laws
BT2 economic theories

ECUADOR
BT1 south america
BT2 america
rt lafta

EDIBLE FUNGI
BT1 vegetables

EDUCATION
NT1 agricultural education
NT1 continuing education
NT1 literacy
NT1 schools
NT2 primary education
NT2 secondary education
NT1 specialized higher education
NT1 universities
NT1 vocational training
NT2 apprenticeship
NT2 on farm training
NT2 practical education
NT2 retraining
rt curriculum
rt leisure
rt teachers

EDUCATIONAL GRANTS
uf scholarship

EDUCATIONAL GRANTS (C)
BT1 educational policy

EDUCATIONAL POLICY
NT1 apprenticeship articles
NT1 educational grants
NT1 educational reform
NT1 study leave
NT1 training levies
rt policy

EDUCATIONAL RADIO
BT1 audiovisual aids
BT2 teaching methods

EDUCATIONAL REFORM
BT1 educational policy

EDUCATIONAL TELEVISION
BT1 audiovisual aids
BT2 teaching methods

eec
USE **european communities**

EFFICIENCY
NT1 utilization of capacity
rt evaluation
rt profit and loss analysis

effluent disposal
USE **waste disposal**

EFTA
(european free trade association)
BT1 organizations
rt austria
rt finland
rt iceland
rt liechtenstein
rt norway
rt portugal
rt sweden
rt switzerland

EGG PRODUCTS
rt eggs
rt food

EGGS
rt egg products
rt food

EGYPT
uf arab republic of egypt
uf united arab republic
BT1 arab countries
BT1 mediterranean countries
BT1 north africa
BT2 africa
rt english-speaking africa

eire
USE **ireland**

EL SALVADOR
uf salvador
BT1 central america
BT2 america
rt cacm

ELASTICITIES
BT1 single equation analysis
BT2 parametric methods
BT3 econometrics
rt price elasticity
rt substitution

ELDERLY FAMILY-WORKERS
BT1 agricultural manpower

ELECTRICITY
BT1 energy
BT2 economic resources
BT3 economic situation
rt electrification
rt fuels

ELECTRIFICATION
rt electricity
rt farm equipment

ELECTRONIC DATA PROCESSING
rt boolean algebra
rt computer software

ELITES
BT1 status
BT2 social stratification
BT3 social structure

EMBARGOS
BT1 discrimination
BT2 protection

EMIGRATION
BT1 migration

EMPLOYMENT
NT1 occupational change
NT1 off-farm employment
NT1 underemployment
NT1 unemployment
NT2 seasonal unemployment
NT2 structural unemployment
rt labour
rt labour economics

ENDIVES
BT1 vegetables

ENERGY
BT1 economic resources
BT2 economic situation
NT1 electricity
NT1 solar energy
rt fuels

ENGLISH-SPEAKING AFRICA
rt africa
rt botswana
rt egypt
rt french-speaking africa
rt gambia
rt ghana
rt kenya
rt lesotho
rt liberia
rt libya
rt malawi
rt mauritius
rt namibia
rt nigeria
rt rhodesia
rt sierra leone
rt south africa
rt st helena
rt sudan
rt swaziland
rt tanzania
rt uganda
rt zambia

ENTERTAINMENT
BT1 leisure

ENVIRONMENT
NT1 ecology
NT1 natural disasters
NT1 rural environment
NT2 agricultural geography
NT2 rural areas
NT2 rural communities
NT3 villages
NT2 rural development
NT3 community development
NT4 rural animation
NT3 settlement
NT1 rural-urban relations
NT2 urbanization
NT1 urban environment
NT2 urban areas
NT3 suburban agriculture
NT3 towns
NT3 urban hinterland
rt environmental policy

ENVIRONMENT (C)
rt landscape
rt pollution by agriculture
rt pollution of agriculture

ENVIRONMENTAL POLICY
NT1 landscape conservation
NT1 nature conservancy
NT1 pollution by agriculture
NT2 agricultural wastes
NT3 crop residues
NT3 livestock wastes
NT2 waste disposal
NT1 pollution control
NT1 pollution of agriculture
NT2 air pollution
NT2 water pollution
rt environment
rt forest policy
rt physical planning
rt policy

EQUATORIAL GUINEA
uf *fernando po*
uf *rio muni*
uf *spanish guinea*
BT1 central africa
BT2 africa south of sahara
BT3 africa

EQUILIBRIUM THEORY
sf *theory*
BT1 economic theories

erosion
USE **soil conservation**

ESTIMATED COSTS
BT1 production costs
BT2 costs
BT3 cost analysis

ETHIOPIA
uf *abyssinia*
BT1 east africa
BT2 africa south of sahara
BT3 africa

ETHNIC GROUPS
BT1 groups
BT1 social anthropology

ethyl acohol
USE **alcohol**

EUROPE
NT1 eastern europe
NT2 albania
NT2 bulgaria
NT2 byelorussian ssr
NT2 czechoslovakia
NT2 germany dr
NT2 hungary
NT2 poland
NT2 romania
NT2 ukrainian ssr
NT2 ussr
NT2 yugoslavia
NT1 western europe
NT2 andorra
NT2 austria
NT2 azores
NT2 belgium
NT3 belgium (antwerp)
NT3 belgium (brabant)
NT3 belgium (east flanders)
NT3 belgium (hainault)
NT3 belgium (liege)
NT3 belgium (limbourg)
NT3 belgium (namur)
NT3 belgium (west flanders)
NT2 canary islands
NT2 cyprus
NT2 france
NT3 france (overseas depts and terr)
NT3 france (r. alsace)
NT3 france (r. aquitaine)
NT3 france (r. auvergne)

EUROPE (C)
NT3 france (r. brittany)
NT3 france (r. burgundy)
NT3 france (r. centre)
NT3 france (r. champagne ardennes)
NT3 france (r. corsica)
NT3 france (r. franche comte)
NT3 france (r. ile de france)
NT3 france (r. languedoc roussillon)
NT3 france (r. limousin)
NT3 france (r. lorraine)
NT3 france (r. lower normandy)
NT3 france (r. midi pyrenees)
NT3 france (r. nord pas de calais)
NT3 france (r. pays de la loire)
NT3 france (r. picardy)
NT3 france (r. poitou charentes)
NT3 france (r. provence alpes cote d'az
NT3 france (r. rhone alpes)
NT3 france (r. upper normandy)
NT2 germany fr
NT3 germany fr (baden-wuerttemberg)
NT3 germany fr (bavaria)
NT3 germany fr (hesse)
NT3 germany fr (lower saxony)
NT3 germany fr (north rhine westphalia)
NT3 germany fr (rhineland palatinate)
NT3 germany fr (saarland)
NT3 germany fr (schleswig-holstein)
NT2 gibraltar
NT2 greece
NT2 ireland
NT3 ireland (connacht)
NT3 ireland (leinster)
NT3 ireland (munster)
NT2 italy
NT3 italy (abruzzi)
NT3 italy (apulia)
NT3 italy (basilicata)
NT3 italy (calabria)
NT3 italy (campania)
NT3 italy (emilia-romagna)
NT3 italy (friuli-venezia giulia)
NT3 italy (latium)
NT3 italy (liguria)
NT3 italy (lombardy)
NT3 italy (marches)
NT3 italy (molise)
NT3 italy (piedmont)
NT3 italy (sardinia)
NT3 italy (sicily)
NT3 italy (trentino-alto adige)
NT3 italy (tuscany)
NT3 italy (umbria)
NT3 italy (valle d'aosta)
NT3 italy (veneto)
NT2 liechtenstein
NT2 luxembourg
NT2 madeira
NT2 malta
NT2 monaco
NT2 netherlands
NT3 netherlands (drenthe)
NT3 netherlands (friesland)
NT3 netherlands (gelderland)
NT3 netherlands (groningen)
NT3 netherlands (n.brabant)
NT3 netherlands (n.holland)
NT3 netherlands (overijssel)
NT3 netherlands (s.holland)
NT3 netherlands (s.limburg)
NT3 netherlands (utrecht)
NT3 netherlands (zeeland)
NT2 portugal
NT2 san marino
NT2 scandinavia
NT3 denmark
NT4 denmark (bornholm)
NT4 denmark (fyn)
NT4 denmark (jutland)
NT4 denmark (lolland-falster)
NT4 denmark (zealand)
NT3 faroe islands
NT3 finland
NT3 greenland
NT3 iceland
NT3 norway
NT3 sweden

EUROPE (C)
NT2 spain
NT2 switzerland
NT2 turkey
NT2 united kingdom
NT3 united kingdom (channel islands)
NT3 united kingdom (england and wales)
NT4 united kingdom (e.england)
NT4 united kingdom (e.midlands)
NT4 united kingdom (n.england)
NT4 united kingdom (s.w.england)
NT4 united kingdom (w.midlands)
NT4 united kingdom (wales)
NT4 united kingdom (yorks/lancs)
NT3 united kingdom (isle of man)
NT3 united kingdom (n.ireland)
NT3 united kingdom (scotland)
NT4 united kingdom (e.scotland)
NT4 united kingdom (highlands+islands)
NT4 united kingdom (n.scotland)
NT4 united kingdom (w.scotland)
NT2 vatican
rt european communities

EUROPEAN COMMUNITIES
uf *eec*
uf *european economic community*
BT1 organizations
rt belgium
rt cap
rt denmark
rt europe
rt france
rt germany fr
rt ireland
rt italy
rt luxembourg
rt netherlands
rt united kingdom

european economic community
USE **european communities**

EUROPEAN FARM ACCOUNTING NETWORK
BT1 farm surveys
BT2 surveys
rt farm income

EUROPEAN PARLIAMENT
BT1 organizations
rt parliament

EUROPEAN REGIONAL DEVELOPMENT FUND
BT1 funds

EUROPEAN SOCIAL FUND
BT1 funds

EUROPEAN UNITS OF ACCOUNT
BT1 currency

EVALUATION
NT1 measurement
rt efficiency
rt planning

EVALUATION OF RESEARCH
BT1 research policy

EVAPORATED MILK
sf *condensed milk*
BT1 milk products

EWES
BT1 sheep

exchange rates
USE **monetary parity**

EXISE DUTIES
BT1 indirect taxation
BT2 fiscal policy
BT3 public finance

EXPERIMENTAL STATIONS
BT1 research institutions
BT2 organization of research

EXPORT CONTROLS
BT1 trade barriers
BT2 protection

EXPORT CREDITS
BT1 credit
BT2 credit transactions
BT1 export promotion
BT2 trade policy

EXPORT DIVERSIFICATION
BT1 trade policy

export guaranties
USE **export insurance**

EXPORT INSURANCE
uf *export guaranties*
BT1 insurances

EXPORT PROMOTION
BT1 trade policy
NT1 export credits
NT1 export refunds
NT1 export subsidies

EXPORT REFUNDS
BT1 export promotion
BT2 trade policy

EXPORT SUBSIDIES
BT1 export promotion
BT2 trade policy

EXPORTS
BT1 external trade
BT2 trade

EXTENDED FAMILIES
BT1 family structure
NT1 multi-generation families

EXTENSION
uf *supervised credit+*
NT1 advisory centres
NT1 advisory officers
rt agricultural press
rt communication
rt diffusion of research
rt information diffusion
rt mass media

EXTENSIVE FARMING
BT1 agricultural systems

EXTENSIVE LIVESTOCK FARMING
uf *ranching*
BT1 livestock farming
BT2 livestock sector

EXTENSIVE PRODUCTION
BT1 fattening
BT2 feeding
BT3 livestock sector

EXTERNAL DEBT
BT1 debt
BT2 credit transactions
rt balance of payments

EXTERNAL TRADE
BT1 trade
NT1 agricultural trade
NT1 exports
NT1 imports

FACTORIAL ANALYSIS
BT1 statistical inference
BT2 statistical analysis
BT3 statistics
rt component analysis

FACTORS OF PRODUCTION
BT1 microeconomic analysis
rt capital
rt labour
rt land

faeroe islands
USE **faroe islands**

FALKLAND ISLANDS
(and dependencies)
uf *south sandwich islands*
BT1 south america
BT2 america
rt united kingdom

FALLOW
BT1 land

FAMILIES
NT1 family and environment
NT1 family size
NT2 children
NT3 daughters
NT3 sons
NT2 grandparents
NT2 kinship
NT2 parents
NT3 fathers
NT3 mothers
NT2 youth
NT3 rural youth
NT1 farm families
NT1 marriage
NT2 divorce
NT2 marriage age
NT1 sexual behaviour
rt households
rt small groups

FAMILY ALLOWANCES
BT1 social security

FAMILY AND ENVIRONMENT
BT1 families

FAMILY BREAKDOWN
BT1 family life

FAMILY FARMS
BT1 types of farming
BT2 agricultural systems
rt medium sized farms

FAMILY INTEGRATION
BT1 family life

FAMILY LABOUR
BT1 agricultural manpower

FAMILY LIFE
NT1 family breakdown
NT1 family integration
NT1 generation conflict
NT2 parent child relationships
NT1 woman's status

FAMILY PLANNING
uf *birth control*
uf *population control*
BT1 population change
rt sexual behaviour

FAMILY SIZE
BT1 families
NT1 children
NT2 daughters
NT2 sons
NT1 grandparents
NT1 kinship
NT1 parents
NT2 fathers
NT2 mothers
NT1 youth
NT2 rural youth
rt family structure

FAMILY STRUCTURE
NT1 extended families
NT2 multi-generation families
NT1 heads of families
NT1 matriarchy
NT1 nuclear families
NT2 one generation families

FAMILY STRUCTURE (C)
NT2 two generation families
NT1 partnerships
NT1 patriarchy
rt family size
rt structure

FAO
(food and agriculture organization)
BT1 organizations

far east
USE **east asia**

FARM ACCOUNTS
BT1 economic accounts

FARM AMALGAMATION
BT1 concentration
BT2 socioeconomic organization

FARM AREA
uf *acreage*
NT1 arable land
NT1 farmland
rt farm size

FARM BUDGETING
sf *budgeting*
BT1 farm planning
BT1 non-optimizing methods
BT2 operations research

FARM BUILDINGS
sf *buildings*
NT1 livestock housing
rt farm inputs
rt fixed capital

farm butter
USE **butter**

FARM CLOSURES
rt severance allowance
rt structural policy

FARM DWELLINGS
sf *buildings*
BT1 rural housing
BT2 living conditions

FARM ENTERPRISES
(branches of farming)
NT1 ancillary enterprises
NT1 crop enterprises
NT2 field crops
NT2 forestry
NT2 grazing
NT2 horticulture
NT3 floriculture
NT3 fruit growing
NT3 tree nurseries
NT3 vegetable growing
NT2 ley farming
NT2 viticulture
NT1 food production
NT1 livestock enterprises
NT2 apiculture
NT2 aquaculture
NT2 cattle farming
NT3 dairy farming
NT2 fish farming
NT2 goat keeping
NT2 horse breeding
NT2 pig farming
NT2 poultry farming
NT2 sheep farming
NT2 silk worms
NT2 small animal rearing
NT1 mixed cropping
rt farms
rt fishing
rt hunting

FARM EQUIPMENT
NT1 farm machinery
NT2 harvester
NT3 combine harvester

FARM EQUIPMENT (C)
NT2 tractors
NT1 tools
rt electrification
rt farm inputs
rt fixed capital
rt mechanization

FARM FAMILIES
BT1 families

FARM HELPER SERVICES
sf *mutual help*
BT1 co-operative services
NT1 co-operative feedstores
NT1 machinery co-operatives
NT1 work banks
rt horizontal integration

FARM HOLIDAYS
BT1 tourism

FARM INCOME
BT1 farm results
rt european farm accounting network

FARM INDEBTEDNESS
BT1 agricultural situation
BT1 farm results
NT1 bankruptcy
rt debt

FARM INPUTS
uf *production materials*
rt farm buildings
rt farm equipment
rt feedingstuffs
rt fertilizers
rt fuels
rt pesticides
rt plants
rt seeds
rt selective herbicides

FARM LEASES
uf *tenancy agreements*
BT1 agricultural tenure

FARM MACHINERY
uf *machinery*
BT1 farm equipment
NT1 harvester
NT2 combine harvester
NT1 tractors

FARM MANAGEMENT
BT1 management
BT2 human resources
BT3 economic resources
BT4 economic situation

FARM MODELS
BT1 farm planning
BT1 models

FARM PLANNING
NT1 farm budgeting
NT1 farm models
NT1 financial planning
NT2 budget accounting
NT1 gross margins analysis
NT1 investment planning
NT2 risk
NT2 uncertainty
NT1 management by objectives
rt accounting
rt agricultural planning
rt farm plans
rt planning

FARM PLANS
BT1 investment promotion
BT2 investment policy
NT1 investment support thresholds
rt farm planning
rt structural policy

FARM PRODUCED FODDER
BT1 feedingstuffs
NT1 dry feed
NT1 green fodder
NT1 hay
NT1 roughage
NT1 silage
NT1 straw

FARM PROFIT
BT1 farm results

farm recreations
USE **tourism**

FARM RESULTS
sf *results*
NT1 earned income
NT1 farm income
NT1 farm indebtedness
NT2 bankruptcy
NT1 farm profit
NT1 farmers income
NT2 non-farm income
NT1 gross returns
NT1 losses
NT1 net returns
rt accounting

FARM ROADS
BT1 roads

FARM SECTOR
sf *agricultural sector*
BT1 primary sector
BT2 economic sectors

FARM SIZE
NT1 livestock numbers
NT1 parcel size
rt farm area
rt farm structure
rt large farms
rt medium sized farms
rt small farms

FARM STRUCTURE
BT1 agricultural structure
rt agricultural systems
rt farm size
rt structure

FARM SURVEYS
BT1 surveys
NT1 european farm accounting network

FARM WORKERS
BT1 occupations
rt agricultural manpower
rt hired labour

FARMERS
BT1 occupations
rt managers

FARMERS ASSOCIATIONS
sf *associations*
BT1 interest groups
BT2 groups

FARMERS ATTITUDES
BT1 economic sociology
rt behaviour

FARMERS INCOME
BT1 farm results
NT1 non-farm income

FARMLAND
BT1 farm area

FARMS
NT1 collective farms
NT2 kibbutzim
NT2 kolkhozy
NT2 moshavim
NT1 private farms
NT1 state farms

FARMS (C)
rt farm enterprises
rt types of farming

FAROE ISLANDS
uf *faeroe islands*
BT1 scandinavia
BT2 western europe
BT3 europe
rt denmark

FAT PRODUCTS
BT1 oils and fats
rt margarine

FATHERS
BT1 parents
BT2 family size
BT3 families

fats
USE **oils and fats**

fatstock
USE **livestock**

FATTENING
(subterms apply to livestock feeding generally to accomodate language equivalents)
BT1 feeding
BT2 livestock sector
NT1 extensive production
NT1 indoor fattening
NT1 intensive fattening
rt controlled grazing

federal legislations
USE **legislation**

FEDERALISM
BT1 political systems

FEED CAKE
BT1 manufactured feeds
BT2 feedingstuffs

FEED CROPS
uf *fodder crops*
uf *forage crops*
BT1 crops
NT1 clover
NT1 lucerne
rt feed grains
rt feed potatoes
rt feedingstuffs
rt fodder beet
rt forage peas
rt ley farming

FEED GRAINS
BT1 cereals
BT2 crops
NT1 coarse grains
rt feed crops
rt ley farming

FEED INDUSTRY
BT1 input industries

FEED POTATOES
BT1 potatoes
BT2 root crops
BT3 crops
rt feed crops

FEED RATIONS
BT1 feeding
BT2 livestock sector

FEED REQUIREMENTS
BT1 feeding
BT2 livestock sector

FEED-MIXES
BT1 manufactured feeds
BT2 feedingstuffs

feeder cattle
USE **beef cattle**

FEEDING
BT1 livestock sector
NT1 fattening
NT2 extensive production
NT2 indoor fattening
NT2 intensive fattening
NT1 feed rations
NT1 feed requirements
NT1 fodder conversion
rt feedingstuffs

FEEDINGSTUFFS
NT1 farm produced fodder
NT2 dry feed
NT2 green fodder
NT2 hay
NT2 roughage
NT2 silage
NT2 straw
NT1 manufactured feeds
NT2 concentrates
NT2 feed cake
NT2 feed-mixes
rt additives
rt farm inputs
rt feed crops
rt feeding

FEMALE LABOUR
BT1 hired labour
BT2 agricultural manpower
rt women

fermented milk
USE **cultured milks**

fermented tea
USE **tea**

fernando po
USE **equatorial guinea**

fertility
USE **soil fertility**

FERTILIZER APPLICATION
BT1 land improvement
BT2 cultivation methods

FERTILIZER INDUSTRY
BT1 chemical industry
BT2 input industries

FERTILIZERS
NT1 inorganic fertilizers
NT2 compost
NT2 compound fertilizers
NT2 lime
NT2 nitrogen
NT3 urea
NT2 phosphates
NT2 potash
NT1 organic fertilizers
NT2 green manure
NT2 guano
NT2 manure
NT2 slurry
rt farm inputs

FEUDALISM
BT1 political systems

FIBRE CROPS
BT1 crops
NT1 cotton
NT1 fibre flax
NT1 hemp
NT1 kapok
NT1 sisal
rt industrial crops

FIBRE FLAX
BT1 fibre crops
BT2 crops

FIELD CROPS
BT1 crop enterprises
BT2 farm enterprises
rt crops

FIGS
BT1 fruits

FIJI
BT1 oceania
rt commonwealth

FILMS
BT1 mass media
rt audiovisual aids

FINAL CONSUMPTION
BT1 aggregate data

FINAL PRODUCTION
BT1 aggregate data

FINANCE
NT1 investment income
NT1 outside finance
NT1 savings
NT1 self finance

FINANCIAL INSTITUTIONS
sf *institutions*
BT1 economic units
rt banks

FINANCIAL MATHEMATICS
BT1 mathematics
NT1 actuarial mathematics

FINANCIAL PLANNING
BT1 farm planning
NT1 budget accounting

FINANCING RESEARCH
BT1 research policy
rt research contracts

FINLAND
BT1 scandinavia
BT2 western europe
BT3 europe
rt efta

FIRE INSURANCE
BT1 agricultural insurance
BT2 insurances

FIRES
BT1 agricultural disasters
NT1 forest fires

FIRMS
NT1 conglomerates
NT1 foreign firms
NT1 multinational corporations
NT1 nationalized enterprises
NT1 private firms
NT2 companies
NT2 private companies
NT1 public enterprises

fiscal measures
USE **fiscal policy**

FISCAL POLICY
uf *fiscal measures*
uf *taxation policy*
BT1 public finance
NT1 direct taxation
NT2 capital taxation
NT3 land tax
NT2 company tax
NT2 income tax
NT2 inheritance tax
NT1 indirect taxation
NT2 exise duties
NT2 stamp duty
NT2 value added tax
NT1 taxes
rt income transfers

FISCAL POLICY (C)
rt levies

FISH
rt fish farming
rt fish products

FISH FARMING
BT1 livestock enterprises
BT2 farm enterprises
rt fish
rt fishing

FISH INDUSTRY
BT1 food industry

FISH PRODUCTS
rt fish
rt food
rt prepared foods

FISHING
rt farm enterprises
rt fish farming
rt leisure

FIXED CAPITAL
uf basic funds
BT1 assets
BT2 balance sheets
BT3 balance sheet analysis
rt capital
rt farm buildings
rt farm equipment

FIXED COSTS
BT1 production costs
BT2 costs
BT3 cost analysis

FLAVOURED MILKS
BT1 milk products

FLAVOURINGS
BT1 additives

FLEXIBILITY
BT1 single equation analysis
BT2 parametric methods
BT3 econometrics

flocks
USE **livestock numbers**

FLOODING
BT1 agricultural disasters

FLORICULTURE
BT1 horticulture
BT2 crop enterprises
BT3 farm enterprises

FLOUR
(only for cereals)
BT1 cereal products
NT1 cornflour
NT1 oatmeal
NT1 rice flour
NT1 rye flour
NT1 wheat flour

FLOWERS
BT1 ornamental plants
BT2 crops
NT1 bulbs

FLUCTUATIONS
BT1 trade cycles
BT2 economic development
NT1 cyclic fluctuations
NT1 seasonal fluctuations
NT1 trends

FOALS
BT1 horses

FODDER BEET
BT1 root crops

FODDER BEET (C)
BT2 crops
rt feed crops

FODDER CONVERSION
BT1 feeding
BT2 livestock sector

fodder crops
USE **feed crops**

FOOD
BT1 products
rt additives
rt bakery products
rt beverages
rt cereal products
rt chocolate
rt cocoa
rt egg products
rt eggs
rt fish products
rt fruit products
rt honey
rt meat
rt meat products
rt milk
rt milk products
rt prepared foods
rt sugar
rt sugar confectionery
rt vegetable products

FOOD AID
sf aid
BT1 development aid
rt food policy

food and agricultural sector
USE **agro-industrial sector**

food balance
USE **food supply**

FOOD GRAINS
BT1 cereals
BT2 crops

FOOD INDUSTRY
NT1 bakery industry
NT2 pasta industry
NT1 beverage industry
NT2 brewing industry
NT2 coffee industry
NT2 distilling industry
NT2 tea industry
NT2 wine industry
NT1 canning industry
NT1 catering industry
NT1 cocoa industry
NT1 confectionery industry
NT1 dairy industry
NT1 fish industry
NT1 frozen foods industry
NT1 meat and livestock industry
NT2 abattoirs
NT1 milling industry
NT1 oils and fats industry
NT1 starch industry
NT1 sugar industry
rt industry
rt post-agricultural sector
rt processing

FOOD INSPECTION
BT1 public health legislation
rt consumer protection
rt quality controls

FOOD POLICY
rt agricultural policy
rt food aid
rt food production
rt food supply
rt hunger
rt nutrition
rt policy

FOOD PRICES
BT1 prices
rt consumer prices

FOOD PRODUCTION
BT1 farm enterprises
rt food policy

FOOD SUPPLY
(includes nutrition statistics)
uf food balance
BT1 supply balance
rt food policy
rt nutrition

forage crops
USE **feed crops**

FORAGE PEAS
BT1 peas
BT2 legumes
BT3 vegetables
rt feed crops

FORECASTS
NT1 projections

FOREIGN EXCHANGE
BT1 currency

FOREIGN FIRMS
BT1 firms

FOREIGN INVESTMENT
BT1 investment
BT2 agricultural situation

foreign workers
USE **migrant labour**

FOREST FIRES
BT1 fires
BT2 agricultural disasters

FOREST LAND
BT1 land
rt forest management
rt forest policy
rt forest products
rt forestry
rt forestry groups
rt national forests
rt private forests
rt public forests

FOREST MANAGEMENT
BT1 land use systems
NT1 afforestation
rt forest land

FOREST OWNERSHIP
NT1 national forests
NT1 private forests
NT1 public forests

FOREST POLICY
rt agricultural policy
rt environmental policy
rt forest land
rt land use systems
rt policy

FOREST PRODUCTS
uf timbers
BT1 primary products
BT2 products
rt forest land
rt forestry
rt paper industry

FORESTRY
BT1 crop enterprises
BT2 farm enterprises
rt forest land
rt forest products

FORESTRY GROUPS
BT1 co-operative farm enterprises

FORESTRY GROUPS (C)
rt forest land

FRANCE
BT1 mediterranean countries
BT1 western europe
BT2 europe
NT1 france (overseas depts and terr)
NT1 france (r. alsace)
NT1 france (r. aquitaine)
NT1 france (r. auvergne)
NT1 france (r. brittany)
NT1 france (r. burgundy)
NT1 france (r. centre)
NT1 france (r. champagne ardennes)
NT1 france (r. corsica)
NT1 france (r. franche comte)
NT1 france (r. ile de france)
NT1 france (r. languedoc roussillon)
NT1 france (r. limousin)
NT1 france (r. lorraine)
NT1 france (r. lower normandy)
NT1 france (r. midi pyrenees)
NT1 france (r. nord pas de calais)
NT1 france (r. pays de la loire)
NT1 france (r. picardy)
NT1 france (r. poitou charentes)
NT1 france (r. provence alpes cote d'az
NT1 france (r. rhone alpes)
NT1 france (r. upper normandy)
rt afars and issas territory
rt andorra
rt antarctica
rt comoro islands
rt european communities
rt french guiana
rt guadeloupe
rt martinique
rt pacific islands fr
rt reunion
rt st pierre and miquelon

FRANCE (OVERSEAS DEPTS AND TERR)
BT1 france
BT2 mediterranean countries
BT2 western europe
BT3 europe

FRANCE (R: ALSACE)
BT1 france
BT2 mediterranean countries
BT2 western europe
BT3 europe

FRANCE (R. AQUITAINE)
BT1 france
BT2 mediterranean countries
BT2 western europe
BT3 europe

FRANCE (R. AUVERGNE)
BT1 france
BT2 mediterranean countries
BT2 western europe
BT3 europe

FRANCE (R. BRITTANY)
BT1 france
BT2 mediterranean countries
BT2 western europe
BT3 europe

FRANCE (R. BURGUNDY)
BT1 france
BT2 mediterranean countries
BT2 western europe
BT3 europe

FRANCE (R. CENTRE)
BT1 france
BT2 mediterranean countries
BT2 western europe
BT3 europe

FRANCE (R. CHAMPAGNE ARDENNES)
BT1 france
BT2 mediterranean countries
BT2 western europe

FRANCE (R. CHAMPAGNE ARDENNES) (C)
BT3 europe

FRANCE (R. CORSICA)
BT1 france
BT2 mediterranean countries
BT2 western europe
BT3 europe

FRANCE (R. FRANCHE COMTE)
BT1 france
BT2 mediterranean countries
BT2 western europe
BT3 europe

FRANCE (R. ILE DE FRANCE)
BT1 france
BT2 mediterranean countries
BT2 western europe
BT3 europe

FRANCE (R. LANGUEDOC ROUSSILLON)
BT1 france
BT2 mediterranean countries
BT2 western europe
BT3 europe

FRANCE (R. LIMOUSIN)
BT1 france
BT2 mediterranean countries
BT2 western europe
BT3 europe

FRANCE (R. LORRAINE)
BT1 france
BT2 mediterranean countries
BT2 western europe
BT3 europe

FRANCE (R. LOWER NORMANDY)
BT1 france
BT2 mediterranean countries
BT2 western europe
BT3 europe

FRANCE (R. MIDI PYRENEES)
BT1 france
BT2 mediterranean countries
BT2 western europe
BT3 europe

FRANCE (R. NORD PAS DE CALAIS)
BT1 france
BT2 mediterranean countries
BT2 western europe
BT3 europe

FRANCE (R. PAYS DE LA LOIRE)
BT1 france
BT2 mediterranean countries
BT2 western europe
BT3 europe

FRANCE (R. PICARDY)
BT1 france
BT2 mediterranean countries
BT2 western europe
BT3 europe

FRANCE (R. POITOU CHARENTES)
BT1 france
BT2 mediterranean countries
BT2 western europe
BT3 europe

FRANCE (R. PROVENCE ALPES COTE D'AZ
(provence alpes cote d'azur)
BT1 france
BT2 mediterranean countries
BT2 western europe
BT3 europe

FRANCE (R. RHONE ALPES)
BT1 france
BT2 mediterranean countries
BT2 western europe
BT3 europe

FRANCE (R. UPPER NORMANDY)
BT1 france
BT2 mediterranean countries
BT2 western europe
BT3 europe

free trade policy
USE **trade liberalization**

freeze-dried coffee
USE **instant coffee**

FREEZE-DRIED FOODS
BT1 dried foods
BT2 prepared foods

FREEZING
BT1 preservation

FRENCH BEANS
BT1 beans
BT2 legumes
BT3 vegetables

FRENCH GUIANA
BT1 south america
BT2 america
rt france

FRENCH POLYNESIA
sf polynesia
BT1 pacific islands fr
BT2 oceania

french somaliland
USE **afars and issas territory**

FRENCH-SPEAKING AFRICA
rt afars and issas territory
rt africa
rt algeria
rt benin
rt burundi
rt cameroon
rt central african empire
rt chad
rt comoro islands
rt congo
rt english-speaking africa
rt gabon
rt guinea
rt ivory coast
rt madagascar
rt mali
rt mauritania
rt morocco
rt niger
rt reunion
rt rwanda
rt senegal
rt togo
rt tunisia
rt upper volta
rt zaire

FRESH FRUITS
BT1 fruits

FRESH VEGETABLES
BT1 vegetables

FROST
BT1 climatic factors

FROZEN FOODS INDUSTRY
BT1 food industry

FROZEN PRICES
BT1 prices

FRUIT GROWING
BT1 horticulture
BT2 crop enterprises
BT3 farm enterprises
rt orchards

FRUIT JUICES
uf fruit squashes

FRUIT JUICES (C)
uf *orange juice*
uf *pineapple juice*
BT1 fruit products
rt soft drinks

FRUIT PRODUCTS
NT1 dried fruit
NT2 prunes
NT2 raisins
NT1 fruit juices
NT1 jams
NT1 marmalades
NT1 nut products
rt food
rt prepared foods

fruit squashes
USE **fruit juices**

FRUITS
NT1 berries
NT2 bilberries
NT2 blackberries
NT2 blackcurrants
NT2 gooseberries
NT2 raspberries
NT2 redcurrants
NT2 strawberries
NT1 citrus fruits
NT2 clementines
NT2 grapefruits
NT2 lemons
NT2 mandarins
NT2 oranges
NT1 figs
NT1 fresh fruits
NT1 grapes
NT1 melons
NT1 persimmons
NT1 pome fruits
NT2 apples
NT2 pears
NT2 quinces
NT1 pomegranates
NT1 rhubarb
NT1 stone fruits
NT2 apricots
NT2 cherries
NT2 olives
NT2 peaches
NT3 nectarines
NT2 plums
NT1 tropical fruits
NT2 avocadoes
NT2 bananas
NT2 dates
NT2 guavas
NT2 mangoes
NT2 passion fruit
NT2 pawpaws
NT2 pineapples
NT1 watermelons
NT1 wild fruits
rt dried fruit
rt nuts

FUELS
rt electricity
rt energy
rt farm inputs

FULL-TIME FARMING
BT1 types of farming
BT2 agricultural systems

FUMIGATION
BT1 plant protection

FUNDS
NT1 eaggf
NT1 european regional development fund
NT1 european social fund
NT1 national fund
rt development aid
rt investment promotion
rt marketing policy
rt regional agricultural policy

FUNDS (C)
rt regional development

FUNGICIDES
BT1 pesticides

further education
USE **continuing education**

FUTURES MARKETS
BT1 commodity exchanges
BT2 marketing channels
BT3 marketing

GABON
BT1 central africa
BT2 africa south of sahara
BT3 africa
rt french-speaking africa

GAEC
(french groups for co-operative farming: groupement agricole d'exploitation en commun)
BT1 co-operative farm enterprises

GAMBIA
BT1 west africa
BT2 africa south of sahara
BT3 africa
rt commonwealth
rt english-speaking africa

GAME
NT1 deer
rt game birds
rt hunting

GAME BIRDS
BT1 poultry
NT1 partridges
NT1 pheasants
NT1 pigeons
rt game

GAME THEORY
BT1 non-optimizing methods
BT2 operations research
rt statistical inference

gardens
USE **domestic gardens**

GARLIC
BT1 vegetables

GATT
(general agreement on tariffs and trade)
BT1 trade negotiations
BT2 trade policy

GEESE
BT1 poultry

GELDINGS
BT1 horses

GENERATION CONFLICT
BT1 family life
NT1 parent child relationships
rt conflict

GENERATIONS
BT1 age

GEOGRAPHICAL MOBILITY
BT1 migration
NT1 commuting
NT1 rural-urban migration
NT1 seasonal migration
NT1 urban rural migration
rt horizontal mobility

GEOGRAPHY
NT1 landscape
NT1 topography
rt agricultural geography
rt economic geography
rt maps

GEOGRAPHY (C)
rt social geography

german federal republic
USE **germany fr**

GERMANY DR
uf *ddr*
uf *east berlin*
uf *eastern germany*
BT1 eastern europe
BT2 europe
rt cmea

GERMANY FR
uf *brd*
uf *german federal republic*
uf *western germany*
BT1 western europe
BT2 europe
NT1 germany fr (baden-wuerttemberg)
NT1 germany fr (bavaria)
NT1 germany fr (hesse)
NT1 germany fr (lower saxony)
NT1 germany fr (north rhine westphalia)
NT1 germany fr (rhineland palatinate)
NT1 germany fr (saarland)
NT1 germany fr (schleswig-holstein)
rt european communities

GERMANY FR (BADEN-WUERTTEMBERG)
BT1 germany fr
BT2 western europe
BT3 europe

GERMANY FR (BAVARIA)
BT1 germany fr
BT2 western europe
BT3 europe

GERMANY FR (HESSE)
BT1 germany fr
BT2 western europe
BT3 europe

GERMANY FR (LOWER SAXONY)
(includes bremen)
BT1 germany fr
BT2 western europe
BT3 europe

GERMANY FR (NORTH RHINE WESTPHALIA)
BT1 germany fr
BT2 western europe
BT3 europe

GERMANY FR (RHINELAND PALATINATE)
BT1 germany fr
BT2 western europe
BT3 europe

GERMANY FR (SAARLAND)
BT1 germany fr
BT2 western europe
BT3 europe

GERMANY FR (SCHLESWIG-HOLSTEIN)
(includes hamburg)
BT1 germany fr
BT2 western europe
BT3 europe

GHANA
BT1 west africa
BT2 africa south of sahara
BT3 africa
rt commonwealth
rt english-speaking africa

GHERKINS
BT1 vegetables

GIBRALTAR
BT1 mediterranean countries
BT1 western europe
BT2 europe
rt united kingdom

GLOBAL ACCOUNTING
BT1 book-keeping systems
BT2 book-keeping

GOAT KEEPING
BT1 livestock enterprises
BT2 farm enterprises
rt goats

GOATS
NT1 billy goats
NT1 kids
NT1 nanny goats
rt goat keeping
rt livestock

GOOSEBERRIES
BT1 berries
BT2 fruits

GOVERNMENT RESEARCH
BT1 research

GRADING
BT1 standards

GRAFTING
BT1 plant husbandry
BT2 cultivation methods

grain
USE **cereals**

GRAIN MAIZE
BT1 maize
BT2 cereals
BT3 crops

grand cayman
USE **cayman islands**

GRANDPARENTS
BT1 family size
BT2 families

grants
SEE **grubbing up grants**
OR **improvement grants**
OR **setting-up grants**
OR **subsidies**
OR **support measures**

GRAPEFRUITS
BT1 citrus fruits
BT2 fruits

GRAPES
BT1 fruits
rt viticulture
rt wine

GRAPHS
BT1 statistics

GRASS
BT1 crops

grasslands
USE **pastures**

GRAZING
BT1 crop enterprises
BT2 farm enterprises
BT1 livestock husbandry
BT2 livestock sector
rt controlled grazing

GRAZING GROUPS
BT1 co-operative farm enterprises

GRAZING TENANCY
BT1 agricultural tenure

GREECE
BT1 mediterranean countries
BT1 western europe
BT2 europe

green belt
USE **open spaces**

GREEN CABBAGES
BT1 cabbages
BT2 vegetables

GREEN FODDER
BT1 farm produced fodder
BT2 feedingstuffs

GREEN MANURE
BT1 organic fertilizers
BT2 fertilizers

GREEN MONEY
BT1 currency

GREEN REVOLUTION
BT1 agricultural development
rt high yielding varieties

GREENHOUSE CROPS
BT1 cropping systems

GREENLAND
BT1 north america
BT2 america
BT1 scandinavia
BT2 western europe
BT3 europe
rt denmark

GRENADA
BT1 west indies associated states
BT2 caribbean
BT3 america

GREY LITERATURE
BT1 publications

GROSS MARGINS
BT1 profitability
BT2 profit and loss analysis

GROSS MARGINS ANALYSIS
BT1 farm planning
rt economic analysis

GROSS NATIONAL PRODUCT
BT1 aggregate data
rt national accounting

GROSS RETURNS
BT1 farm results

GROUND RENT
BT1 valuation

GROUNDNUT OILS
BT1 vegetable oils
BT2 oils and fats

GROUNDNUTS
uf *peanuts*
BT1 oilseeds
BT2 crops

GROUP BEHAVIOUR
BT1 group integration
BT2 groups

GROUP CONSCIOUSNESS
BT1 group integration
BT2 groups

GROUP INTEGRATION
BT1 groups
NT1 group behaviour
NT1 group consciousness
NT1 group interaction
NT1 group solidarity

GROUP INTERACTION
BT1 group integration
BT2 groups

GROUP SIZE
BT1 groups
NT1 large groups
NT1 small groups
NT2 neighbourhood

GROUP SOLIDARITY
BT1 group integration
BT2 groups

GROUPEMENT FONCIER AGRICOLE
(in france only)
BT1 co-operative farm enterprises

GROUPS
NT1 age groups
NT1 clubs
NT1 ethnic groups
NT1 group integration
NT2 group behaviour
NT2 group consciousness
NT2 group interaction
NT2 group solidarity
NT1 group size
NT2 large groups
NT2 small groups
NT3 neighbourhood
NT1 interest groups
NT2 agricultural trade unions
NT2 farmers associations
NT2 pressure groups
NT2 trade associations
NT2 trade unions
NT1 membership
NT2 individuals and groups
NT1 working groups
rt communes

growth
USE **economic growth**

GROWTH RATE
BT1 economic growth
BT2 economic development

GROWTH THEORY
sf *theory*
BT1 economic theories
rt economic growth

GRUBBING UP GRANTS
sf *grants*
BT1 subsidies
BT2 agricultural financial policy

GUADELOUPE
BT1 caribbean
BT2 america
rt france

GUANO
BT1 organic fertilizers
BT2 fertilizers

GUARANTEED MINIMUM PRODUCER PRICES
BT1 agricultural prices
BT2 prices

GUATEMALA
BT1 central america
BT2 america
rt cacm

GUAVAS
BT1 tropical fruits
BT2 fruits

GUIDE PRICES
(cap price for beaf veal and wine)
BT1 agricultural prices
BT2 prices
BT1 prices

GUINEA
BT1 west africa
BT2 africa south of sahara
BT3 africa
rt french-speaking africa

GUINEA-BISSAU
BT1 west africa
BT2 africa south of sahara
BT3 africa

GUYANA
BT1 south america
BT2 america
rt commonwealth

HAIL
BT1 precipitation
BT2 climatic factors

HAIL INSURANCE
BT1 agricultural insurance
BT2 insurances

HAITI
BT1 caribbean
BT2 america
BT1 central america

HAM
(includes cooked and raw ham)
BT1 meat products
rt pigmeat

HAND REARING
BT1 livestock farming
BT2 livestock sector

HANDLING
uf *mechanical handling*
NT1 bulk handling
rt storage
rt transport

HARD CHEESE
BT1 cheese
BT2 milk products

HARD WHEAT
uf *durum wheat*
BT1 wheat
BT2 cereals
BT3 crops

HARDWARE
NT1 computers
NT2 minicomputers

HARROWING
BT1 tillage
BT2 cultivation methods

HARVESTER
BT1 farm machinery
BT2 farm equipment
NT1 combine harvester

HARVESTING
BT1 cultivation methods

hashemite kingdom of jordan
USE **jordan**

HAY
BT1 farm produced fodder
BT2 feedingstuffs

HAZELNUTS
BT1 nuts

HEADS OF FAMILIES
BT1 family structure

HEALTH
BT1 population change
rt nutrition

HEALTH SERVICES
BT1 social security
rt insurances

HEIFERS
BT1 cattle

HEMP
uf *jute*
BT1 fibre crops
BT2 crops

HENS
uf *chickens+*
uf *laying hens*
BT1 broilers
BT2 poultry

HERBICIDES
BT1 pesticides

herbs
SEE **culinary herbs**
OR **medicinal herbs**

HERD IMPROVEMENT
BT1 livestock sector

herds
USE **livestock numbers**

HERITABLE LEASES
BT1 agricultural tenure

HICKORY NUTS
BT1 nuts

high calory foods
USE **prepared foods**

HIGH YIELDING VARIETIES
uf *hyv*
uf *improved varieties*
BT1 varieties
BT2 plant breeding
rt green revolution
rt seeds
rt triticale
rt yields

higher education
SEE **specialized higher education**
OR **universities**

hill areas
USE **mountain areas**

HIRE PURCHASE
BT1 credit transactions

HIRED LABOUR
BT1 agricultural manpower
NT1 female labour
NT1 migrant labour
NT1 part time labour
NT1 permanent labour
NT1 seasonal labour
NT1 skilled labour
NT1 unskilled labour
rt farm workers

HISTORY
rt documentation

holland
USE **netherlands**

holy see
USE **vatican**

home trade
USE **domestic trade**

HONDURAS
BT1 central america
BT2 america
rt cacm

honduras (british)
USE **belize**

HONEY
rt apiculture
rt bees
rt food

HONG KONG
BT1 east asia
BT2 asia
rt united kingdom

HOPS
BT1 crops

HORIZONTAL INTEGRATION
BT1 integration
BT2 socioeconomic organization
rt farm helper services

HORIZONTAL MOBILITY
BT1 social mobility
BT2 social stratification
BT3 social structure
rt geographical mobility

HORSE BEANS
BT1 beans
BT2 legumes
BT3 vegetables

HORSE BREEDING
BT1 livestock enterprises
BT2 farm enterprises
rt horses

HORSE MEAT
BT1 meat

HORSES
NT1 foals
NT1 geldings
NT1 mares
NT1 stallions
rt horse breeding

HORTICULTURE
BT1 crop enterprises
BT2 farm enterprises
NT1 floriculture
NT1 fruit growing
NT1 tree nurseries
NT1 vegetable growing

hourly wages
USE **time wages**

HOUSEHOLD CONSUMPTION
BT1 aggregate data
BT1 consumption
rt consumer behaviour
rt consumer expenditure
rt demand

HOUSEHOLD EXPENDITURE
BT1 consumer expenditure
BT2 aggregate data

HOUSEHOLDS
uf *housewives+*
BT1 economic units
NT1 agricultural households
rt families

housewives+
USE **households**
AND **women**

HUMAN RESOURCES
BT1 economic resources
BT2 economic situation
NT1 labour
NT1 management
NT2 farm management
NT2 self management

HUMID ZONES
BT1 climatic areas

HUNGARY
BT1 eastern europe
BT2 europe
rt cmea

HUNGER
BT1 poverty
BT2 living standards
BT3 living conditions
rt food policy

HUNTING
rt farm enterprises
rt game
rt leisure

HYBRIDIZATION
BT1 plant breeding
rt cross breeding

hybrids
USE **cross breeding**

HYDROPONICS
uf *soilless culture*
BT1 cropping systems

hyv
USE **high yielding varieties**

ibrd
USE **world bank**

ICELAND
BT1 scandinavia
BT2 western europe
BT3 europe
rt efta

IDENTITY
BT1 individuals

IDEOLOGY
BT1 behaviour

imitation foods
USE **simulated foods**

IMMIGRATION
BT1 migration

IMPERFECT COMPETITION
uf *restrictives practices*
BT1 competition
BT2 market economics
NT1 monopoly
NT1 monopsony
NT1 oligopoly
NT1 oligopsony
rt dumping

IMPERIALISM
BT1 political systems
NT1 colonialism

IMPLEMENTATION OF RESEARCH
BT1 research policy

IMPORT CONTROLS
NT1 import levies
NT2 variable levies
rt trade barriers

IMPORT LEVIES
BT1 import controls
BT1 levies
NT1 variable levies

IMPORTS
BT1 external trade
BT2 trade

IMPORTS PROMOTION
BT1 trade policy

improved varieties
USE **high yielding varieties**

IMPROVEMENT GRANTS
sf *grants*
BT1 subsidies
BT2 agricultural financial policy

INCOME
NT1 income distribution
NT2 income transfers
rt accounting
rt national accounting

INCOME ACCOUNTS
BT1 economic accounts

INCOME DISTRIBUTION
BT1 income
NT1 income transfers
rt disparity
rt living standards
rt parity

INCOME TAX
BT1 direct taxation
BT2 fiscal policy
BT3 public finance

INCOME TRANSFERS
BT1 income distribution
BT2 income
rt fiscal policy

indebtedness
USE **debt**

INDEMNIFICATION
(of joint heirs to an indivisible estate)
BT1 succession
BT2 acquisition of ownership
rt indivisible estates
rt partition of estates

INDEXING
BT1 document analysis
BT2 information processing

INDIA
BT1 south east asia
BT2 asia
rt commonwealth

INDICATIVE PLANNING
BT1 planning

INDIRECT TAXATION
sf *taxation*
BT1 fiscal policy
BT2 public finance
NT1 exise duties
NT1 stamp duty
NT1 value added tax
rt tariffs

INDIVIDUALS
NT1 character
NT1 identity
NT1 interpersonal relations
NT1 self esteem
NT1 self perception

INDIVIDUALS AND GROUPS
BT1 membership
BT2 groups

INDIVISIBLE ESTATES
BT1 succession
BT2 acquisition of ownership
rt indemnification
rt non-transferable property

INDOCHINA
BT1 south east asia
BT2 asia
NT1 kampuchea
NT1 lao
NT1 vietnam

INDONESIA
uf *borneo*
BT1 south east asia
BT2 asia

INDOOR FATTENING
BT1 fattening

INDOOR FATTENING (C)
BT2 feeding
BT3 livestock sector

INDUSTRIAL BANKS
BT1 banks
rt development banks
rt investment banks

INDUSTRIAL CROPS
NT1 rubber
NT1 tobacco
rt fibre crops
rt oilseeds

INDUSTRIAL ILLNESS
BT1 safety at work
BT2 working conditions

INDUSTRIAL POTATOES
BT1 potatoes
BT2 root crops
BT3 crops

INDUSTRIAL REARING
BT1 livestock farming
BT2 livestock sector

INDUSTRIAL SOCIETIES
BT1 societies

INDUSTRIALIZATION
BT1 modernization
BT2 social change

INDUSTRY
NT1 rural industry
rt food industry
rt input industries
rt non-food industries

industry and agriculture
USE **agro-industrial relations**

INFANT FOODS
uf *baby foods*
BT1 dietetic foods
BT2 prepared foods

INFLATION
BT1 monetary situation

INFORMATION DIFFUSION
NT1 public relations
rt communication
rt diffusion of research
rt documentation
rt extension
rt information processing
rt mass media

INFORMATION PROCESSING
NT1 document analysis
NT2 abstracting
NT2 classification
NT2 indexing
NT1 documentation languages
NT2 thesauri
NT1 information storage
NT2 microform
rt information diffusion

INFORMATION RETRIEVAL
NT1 card services
NT1 retrospective searches
NT1 sdi services
rt documentation

INFORMATION SERVICES
rt libraries

INFORMATION STORAGE
BT1 information processing
NT1 microform
rt data banks
rt data bases

INFRASTRUCTURE
rt rural planning

INHERITANCE
BT1 property transfers
BT2 land ownership
rt succession

INHERITANCE TAX
BT1 direct taxation
BT2 fiscal policy
BT3 public finance

INLAND TRANSPORT
BT1 transport

INLAND WATERWAYS
BT1 transport

INORGANIC FERTILIZERS
BT1 fertilizers
NT1 compost
NT1 compound fertilizers
NT1 lime
NT1 nitrogen
NT2 urea
NT1 phosphates
NT1 potash

INPUT INDUSTRIES
NT1 agricultural machinery industry
NT1 chemical industry
NT2 fertilizer industry
NT1 feed industry
NT1 seed industry
rt industry
rt pre-agricultural sector
rt processing

INPUT PRICES
BT1 prices

INPUT-OUTPUT ANALYSIS
BT1 non-optimizing methods
BT2 operations research
rt accounting
rt economic analysis

INSECTICIDES
BT1 pesticides

INSTANT COCOA
BT1 instant foods
BT2 prepared foods
rt cocoa
rt dried foods

INSTANT COFFEE
uf coffee powder
uf freeze-dried coffee
uf soluble coffee
BT1 instant foods
BT2 prepared foods
rt coffee
rt dried foods

INSTANT FOODS
BT1 prepared foods
NT1 instant cocoa
NT1 instant coffee
NT1 instant tea

INSTANT TEA
uf tea powder
BT1 instant foods
BT2 prepared foods
rt dried foods
rt tea

INSTITUTION BUILDING
BT1 social institutions
BT2 social systems

institutions
SEE **financial institutions**
OR **organizations**
OR **research institutions**
OR **social institutions**

INSURANCE BENEFITS
BT1 insurances
rt compensation

INSURANCE PREMIUMS
BT1 insurances

INSURANCES
NT1 agricultural insurance
NT2 crop insurance
NT2 fire insurance
NT2 hail insurance
NT2 livestock insurance
NT1 credit insurance
NT1 export insurance
NT1 insurance benefits
NT1 insurance premiums
rt accident benefits
rt health services
rt maternity benefits
rt old age benefits
rt sickness benefits
rt social insurance
rt social security
rt social services
rt unemployment insurance

INTEGER PROGRAMMING
uf mixed integer programming
BT1 programming
rt optimization methods

INTEGRATED PLANT PROTECTION
BT1 plant husbandry
BT2 cultivation methods
rt plant protection

integrated rural development
USE **rural development**

INTEGRATION
BT1 socioeconomic organization
NT1 horizontal integration
NT1 vertical integration
rt co-operation
rt contract legislation
rt diversification

INTENSIVE FARMING
BT1 agricultural systems

INTENSIVE FATTENING
BT1 fattening
BT2 feeding
BT3 livestock sector

inter co-operative enterprises
USE **co-operative farm enterprises**

INTEREST GROUPS
BT1 groups
NT1 agricultural trade unions
NT1 farmers associations
NT1 pressure groups
NT1 trade associations
NT1 trade unions
rt organizations

INTEREST RATES
BT1 credit control
BT2 credit policy
NT1 preferential interest rates

INTEREST REPAYMENTS
sf debt servicing
BT1 repayment
BT2 credit policy

INTERMEDIATE CONSUMPTION
BT1 aggregate data

intermediate technology
USE **appropriate technology**

INTERNATIONAL AGREEMENTS
BT1 trade agreements
BT2 trade policy
rt development aid

INTERNATIONAL LOANS
BT1 loans
BT2 credit transactions

INTERNATIONAL TRADE
uf world trade
BT1 trade

INTERNATIONAL TRADE MODELS
BT1 models
BT1 optimization methods
BT2 operations research

INTERNATIONAL TRANSPORT
BT1 transport

INTERPERSONAL RELATIONS
BT1 individuals

INTERREGIONAL PLANNING
BT1 planning

INTERSECTORAL PLANNING
BT1 planning

INTERVENTION
BT1 market regulations
BT2 marketing policy
BT3 agricultural policy
rt policy

INTERVENTION PRICES
BT1 agricultural prices
BT2 prices

INTRA-COMMUNITY MARKETS
BT1 markets

INTRA-COMMUNITY TRADE
BT1 trade

INVALID BENEFITS
BT1 social insurance
BT2 social security
rt accident prevention

INVENTORIES
BT1 book-keeping

INVESTMENT
BT1 agricultural situation
NT1 foreign investment
NT1 private investment
NT1 public investment
NT1 reinvestment
rt capital

INVESTMENT BANKS
BT1 banks
rt development banks
rt industrial banks
rt world bank

INVESTMENT FUNCTIONS
BT1 single equation analysis
BT2 parametric methods
BT3 econometrics

INVESTMENT GUARANTEES
BT1 investment promotion
BT2 investment policy

INVESTMENT INCOME
BT1 finance

INVESTMENT PLANNING
BT1 farm planning
NT1 risk
NT1 uncertainty

INVESTMENT POLICY
NT1 investment promotion
NT2 farm plans
NT3 investment support thresholds
NT2 investment guarantees
NT1 investment requirements
rt policy

INVESTMENT PROMOTION
BT1 investment policy
NT1 farm plans
NT2 investment support thresholds
NT1 investment guarantees
rt funds

INVESTMENT REQUIREMENTS
BT1 investment policy

INVESTMENT SUPPORT THRESHOLDS
BT1 farm plans
BT2 investment promotion
BT3 investment policy

IRAN
BT1 west asia
BT2 asia

IRAQ
BT1 arab countries
BT1 west asia
BT2 asia

IRELAND
uf eire
uf irish republic
BT1 western europe
BT2 europe
NT1 ireland (connacht)
NT1 ireland (leinster)
NT1 ireland (munster)
rt european communities

IRELAND (CONNACHT)
BT1 ireland
BT2 western europe
BT3 europe

IRELAND (LEINSTER)
BT1 ireland
BT2 western europe
BT3 europe

IRELAND (MUNSTER)
BT1 ireland
BT2 western europe
BT3 europe

irish republic
USE **ireland**

IRRIGATED FARMING
BT1 cropping systems

IRRIGATION
BT1 land improvement
BT2 cultivation methods
rt water

isoglucose
USE **isomerose**

ISOMEROSE
uf isoglucose
uf starch sugar
BT1 sugar
rt artificial sweeteners

ISRAEL
BT1 mediterranean countries
BT1 west asia
BT2 asia

ITALY
BT1 mediterranean countries
BT1 western europe
BT2 europe
NT1 italy (abruzzi)
NT1 italy (apulia)
NT1 italy (basilicata)
NT1 italy (calabria)
NT1 italy (campania)
NT1 italy (emilia-romagna)
NT1 italy (friuli-venezia giulia)
NT1 italy (latium)
NT1 italy (liguria)
NT1 italy (lombardy)

ITALY (C)
NT1 italy (marches)
NT1 italy (molise)
NT1 italy (piedmont)
NT1 italy (sardinia)
NT1 italy (sicily)
NT1 italy (trentino-alto adige)
NT1 italy (tuscany)
NT1 italy (umbria)
NT1 italy (valle d'aosta)
NT1 italy (veneto)
rt european communities

ITALY (ABRUZZI)
BT1 italy
BT2 mediterranean countries
BT2 western europe
BT3 europe

ITALY (APULIA)
BT1 italy
BT2 mediterranean countries
BT2 western europe
BT3 europe

ITALY (BASILICATA)
BT1 italy
BT2 mediterranean countries
BT2 western europe
BT3 europe

ITALY (CALABRIA)
BT1 italy
BT2 mediterranean countries
BT2 western europe
BT3 europe

ITALY (CAMPANIA)
BT1 italy
BT2 mediterranean countries
BT2 western europe
BT3 europe

ITALY (EMILIA-ROMAGNA)
BT1 italy
BT2 mediterranean countries
BT2 western europe
BT3 europe

ITALY (FRIULI-VENEZIA GIULIA)
BT1 italy
BT2 mediterranean countries
BT2 western europe
BT3 europe

ITALY (LATIUM)
BT1 italy
BT2 mediterranean countries
BT2 western europe
BT3 europe

ITALY (LIGURIA)
BT1 italy
BT2 mediterranean countries
BT2 western europe
BT3 europe

ITALY (LOMBARDY)
BT1 italy
BT2 mediterranean countries
BT2 western europe
BT3 europe

ITALY (MARCHES)
BT1 italy
BT2 mediterranean countries
BT2 western europe
BT3 europe

ITALY (MOLISE)
BT1 italy
BT2 mediterranean countries
BT2 western europe
BT3 europe

ITALY (PIEDMONT)
BT1 italy
BT2 mediterranean countries

ITALY (PIEDMONT) (C)
BT2 western europe
BT3 europe

ITALY (SARDINIA)
BT1 italy
BT2 mediterranean countries
BT2 western europe
BT3 europe

ITALY (SICILY)
BT1 italy
BT2 mediterranean countries
BT2 western europe
BT3 europe

ITALY (TRENTINO-ALTO ADIGE)
BT1 italy
BT2 mediterranean countries
BT2 western europe
BT3 europe

ITALY (TUSCANY)
BT1 italy
BT2 mediterranean countries
BT2 western europe
BT3 europe

ITALY (UMBRIA)
BT1 italy
BT2 mediterranean countries
BT2 western europe
BT3 europe

ITALY (VALLE D'AOSTA)
BT1 italy
BT2 mediterranean countries
BT2 western europe
BT3 europe

ITALY (VENETO)
BT1 italy
BT2 mediterranean countries
BT2 western europe
BT3 europe

IVORY COAST
BT1 west africa
BT2 africa south of sahara
BT3 africa
rt french-speaking africa

JAMAICA
BT1 caribbean
BT2 america
BT1 central america
rt commonwealth

JAMS
BT1 fruit products

JAPAN
BT1 east asia
BT2 asia

jars
USE **packs**

job satisfaction
USE **work satisfaction**

JORDAN
uf hashemite kingdom of jordan
BT1 arab countries
BT1 west asia
BT2 asia

jugoslavia
USE **yugoslavia**

JURISPRUDENCE
BT1 case law
rt agricultural tribunals

jute
USE **hemp**

kaffir corn
USE **sorghum**

KAMPUCHEA
uf cambodia
uf khmer republic
BT1 indochina
BT2 south east asia
BT3 asia

KAPOK
BT1 fibre crops
BT2 crops

KENYA
BT1 east africa
BT2 africa south of sahara
BT3 africa
rt commonwealth
rt english-speaking africa

KEYNSIAN ECONOMICS
BT1 economic schools
BT2 economic theories

khmer republic
USE **kampuchea**

KIBBUTZIM
BT1 collective farms
BT2 farms

KIDNEY BEANS
BT1 beans
BT2 legumes
BT3 vegetables

KIDS
BT1 goats

KINSHIP
BT1 family size
BT2 families

kitchen gardens
USE **domestic gardens**

KOHLRABI
BT1 vegetables

KOLKHOZY
BT1 collective farms
BT2 farms

KOREA DPR
uf north korea
BT1 east asia
BT2 asia

KOREA R
uf south korea
BT1 east asia
BT2 asia

KUWAIT
BT1 persian gulf states
BT2 asia

LABELLING
sf brands
BT1 sales promotion
BT2 marketing techniques
NT1 quality labelling
NT2 standard labelling
NT1 trade marks
rt packaging

LABELLING CONTROLS
BT1 quality controls
BT2 consumer protection

LABORATORIES
BT1 research institutions
BT2 organization of research

LABOUR
uf work
BT1 human resources

LABOUR (C)
BT2 economic resources
BT3 economic situation
rt agricultural manpower
rt employment
rt factors of production
rt labour economics

LABOUR ALLOCATION
BT1 organization of work
BT2 labour economics

labour banks
USE **work banks**

LABOUR CONTRACTS
BT1 labour economics
BT1 work planning
BT2 labour economics

LABOUR COSTS
BT1 labour economics

LABOUR ECONOMICS
NT1 division of labour
NT1 labour contracts
NT1 labour costs
NT1 organization of work
NT2 labour allocation
NT1 remuneration
NT2 piece work wages
NT2 time wages
NT2 wage rates
NT2 wages
NT2 wages in kind
NT1 wage agreements
NT1 work planning
NT2 labour contracts
NT2 labour requirements
NT2 work norms
NT1 work study
NT2 rest periods
NT2 unoccupied time
NT2 working hours
rt agricultural manpower
rt agricultural work force
rt employment
rt labour
rt labour legislation
rt leisure
rt working conditions

LABOUR LEGISLATION
BT1 social legislation
NT1 labour protection
NT1 strikes
rt labour economics

LABOUR MARKET
BT1 markets
NT1 labour mobility

LABOUR MOBILITY
sf mobility
BT1 labour market
BT2 markets

labour payment
SEE **remuneration**
OR **wages**

LABOUR PRODUCTIVITY
BT1 productivity
BT2 agricultural situation
BT2 profit and loss analysis

LABOUR PROTECTION
BT1 labour legislation
BT2 social legislation
rt trade unions
rt working conditions

LABOUR REQUIREMENTS
BT1 work planning
BT2 labour economics

LAFTA
(latin american free trade area)

LAFTA (C)
BT1 organizations
rt argentina
rt bolivia
rt brazil
rt chile
rt colombia
rt ecuador
rt mexico
rt paraguay
rt peru
rt uruguay
rt venezuela

LAMB
BT1 meat

LAMBS
BT1 sheep

LAND
NT1 abandoned land
NT1 arable land
NT1 fallow
NT1 forest land
NT1 orchards
NT1 pastures
NT2 permanent pastures
NT2 prairies
NT2 temporary grasslands
NT1 reclaimed land
NT2 polder
NT1 vineyards
NT1 virgin lands
NT1 waste land
NT1 watersheds
rt factors of production
rt land use
rt natural resources

LAND BANKS
rt land diversion
rt safer

LAND CLEARANCE
BT1 cultivation methods

LAND CONSOLIDATION
BT1 structural policy
BT2 agricultural policy
rt land transfers
rt ownership
rt parcel size

land development+
USE **change**
AND **land use**

LAND DIVERSION
BT1 production policy
rt agricultural policy
rt land banks

LAND IMPROVEMENT
uf melioration
BT1 cultivation methods
NT1 drainage
NT1 fertilizer application
NT1 irrigation
NT1 soil conservation
NT1 spray irrigation
rt agricultural engineering

LAND IMPROVMENT CO-OPERATIVES
BT1 co-operative services

LAND MARKETS
BT1 markets
NT1 land mobility
rt land prices

LAND MOBILITY
sf mobility
BT1 land markets
BT2 markets
rt non-transferable property

land mortgage banks
USE **agricultural banks**

LAND OWNERSHIP
NT1 property transfers
NT2 inheritance
NT2 land transfers
NT2 mortgages
rt cadasters
rt ownership
rt ownership conditions

LAND PRICES
BT1 valuation
rt land markets

LAND PRODUCTIVITY
BT1 productivity
BT2 agricultural situation
BT2 profit and loss analysis

LAND REFORM
BT1 agrarian reform

land registers
USE **cadasters**

LAND TAX
BT1 capital taxation
BT2 direct taxation
BT3 fiscal policy
BT4 public finance

land tenure
USE **tenure systems**

LAND TRANSFERS
BT1 property transfers
BT2 land ownership
rt compulsory purchase
rt land consolidation

LAND USE
uf *land development+*
rt cultivation methods
rt land
rt landscape conservation
rt zoning

LAND USE SYSTEMS
NT1 forest management
NT2 afforestation
NT1 public domain
NT2 common lands
NT2 national forests
NT2 national parks
NT1 water management
rt forest policy

landlords
USE **landowners**

LANDLORDS RIGHT
rt landowners

LANDOWNERS
uf *landlords*
rt agricultural tenure
rt landlords right
rt resumption of possession
rt tenants

LANDSCAPE
BT1 geography
rt environment
rt landscape conservation
rt open spaces

LANDSCAPE CONSERVATION
BT1 environmental policy
rt land use
rt landscape

LAO
uf *laos*
BT1 indochina
BT2 south east asia
BT3 asia

laos
USE **lao**

LARGE FARMS
BT1 types of farming
BT2 agricultural systems
rt farm size

LARGE GROUPS
BT1 group size
BT2 groups

LARVICIDES
BT1 pesticides

latin america
SEE **central america**
OR **mexico**
OR **south america**

LAUNCHING A PRODUCT
BT1 sales promotion
BT2 marketing techniques

lauric oils
USE **coconut oils**

LAWS
uf *acts of parliament*
uf *statutes*
BT1 legislation
BT2 legal sources
NT1 subnational legislation

laying hens
USE **hens**

LEADERSHIP
BT1 power
BT2 social systems

LEASES
BT1 tenancy
BT2 tenure systems
rt agricultural tenure

leasing
SEE **capital leasing**
OR **tenancy**

LEAST DEVELOPED COUNTRIES
sf *less developed countries*
BT1 economic geography

LEAST SQUARES METHOD
BT1 statistical inference
BT2 statistical analysis
BT3 statistics

LEATHER INDUSTRY
BT1 non-food industries

LEBANON
BT1 arab countries
BT1 mediterranean countries
BT1 west asia
BT2 asia

LEEKS
BT1 vegetables

LEGAL FORM
uf *company statutes*
rt commercial law
rt companies
rt company law
rt legislation

LEGAL PRINCIPLES
BT1 case law

LEGAL RIGHTS
BT1 legal systems
BT2 social systems

LEGAL SOURCES
NT1 collective agreements
NT1 constitutional law

LEGAL SOURCES (C)
NT1 customary law
NT1 ec laws
NT2 treaty of rome
NT3 ec regulations
NT1 legislation
NT2 civil law
NT2 code forestier
NT2 codified rural legislation
NT2 laws
NT3 subnational legislation
NT2 regulations
NT3 decrees
NT1 local byelaws
NT1 professional codes

LEGAL SYSTEMS
BT1 social systems
NT1 legal rights
rt social security

LEGISLATION
uf *federal legislations*
BT1 legal sources
NT1 civil law
NT1 code forestier
NT1 codified rural legislation
NT1 laws
NT2 subnational legislation
NT1 regulations
NT2 decrees
rt legal form
rt social policy

LEGUMES
BT1 vegetables
NT1 beans
NT2 broad beans
NT2 french beans
NT2 horse beans
NT2 kidney beans
NT2 lima beans
NT2 white beans
NT1 lentils
NT1 peas
NT2 forage peas

LEISURE
uf *recreation*
NT1 entertainment
NT1 sport
rt education
rt fishing
rt hunting
rt labour economics
rt mass media

LEMONS
BT1 citrus fruits
BT2 fruits

LENTILS
BT1 legumes
BT2 vegetables

LESOTHO
BT1 southern africa
BT2 africa south of sahara
BT3 africa
rt commonwealth
rt english-speaking africa

less developed countries
SEE **developing countries**
OR **least developed countries**

LESS FAVOURED AREAS
uf *low income areas*
uf *problem areas*
BT1 regional agricultural policy
BT2 agricultural policy
NT1 mountain areas
rt cap
rt structural policy

LETTUCES
BT1 vegetables

LEVIES
NT1 import levies
NT2 variable levies
rt fiscal policy

LEY FARMING
BT1 crop enterprises
BT2 farm enterprises
rt feed crops
rt feed grains

LIABILITIES
BT1 balance sheets
BT2 balance sheet analysis
NT1 balance adjustment items
NT2 consolidated balance sheets
NT1 outside capital
NT1 owners equity
NT1 reserve funds

liberalism
USE **classical economics**

LIBERIA
BT1 west africa
BT2 africa south of sahara
BT3 africa
rt english-speaking africa

LIBRARIES
rt information services

LIBYA
uf *libyan arab jamahiriya*
uf *libyan arab republic*
BT1 arab countries
BT1 mediterranean countries
BT1 north africa
BT2 africa
rt english-speaking africa

libyan arab jamahiriya
USE **libya**

libyan arab republic
USE **libya**

LICENSES AND PERMITS
NT1 building permits
rt regulations

LIECHTENSTEIN
BT1 western europe
BT2 europe
rt efta

LIFE INSURANCE
BT1 social security

LIMA BEANS
BT1 beans
BT2 legumes
BT3 vegetables

LIME
BT1 inorganic fertilizers
BT2 fertilizers

LINEAR PROGRAMMING
BT1 programming

LINSEED
BT1 oilseeds
BT2 crops

LIQUEURS
BT1 alcoholic beverages
BT2 beverages

LIQUID ASSETS
BT1 working capital
BT2 assets
BT3 balance sheets
BT4 balance sheet analysis

LIQUIDITY
BT1 profit and loss analysis

LITERACY
BT1 education
rt curriculum

LIVESTOCK
uf *fatstock*
uf *slaughter animals*
sf *animals*
rt cattle
rt goats
rt livestock numbers
rt pigs
rt poultry
rt sheep

LIVESTOCK CAPITAL
BT1 assets
BT2 balance sheets
BT3 balance sheet analysis

LIVESTOCK ENTERPRISES
BT1 farm enterprises
NT1 apiculture
NT1 aquaculture
NT1 cattle farming
NT2 dairy farming
NT1 fish farming
NT1 goat keeping
NT1 horse breeding
NT1 pig farming
NT1 poultry farming
NT1 sheep farming
NT1 silk worms
NT1 small animal rearing
rt animal breeding
rt livestock farming

LIVESTOCK FARMING
BT1 livestock sector
NT1 extensive livestock farming
NT1 hand rearing
NT1 industrial rearing
rt livestock enterprises

LIVESTOCK GROUPS
BT1 co-operative farm enterprises

LIVESTOCK HOUSING
sf *buildings*
BT1 farm buildings
rt building controls

LIVESTOCK HUSBANDRY
uf *animal husbandry*
BT1 livestock sector
NT1 controlled grazing
NT1 grazing
NT1 slurry removal
NT1 stabling
NT1 transhumance

LIVESTOCK INSURANCE
BT1 agricultural insurance
BT2 insurances

LIVESTOCK LEASES
BT1 agricultural tenure

LIVESTOCK NUMBERS
uf *flocks*
uf *herds*
BT1 farm size
rt livestock

LIVESTOCK SECTOR
NT1 animal breeding
NT2 artificial insemination
NT2 breeds
NT2 cross breeding
NT2 mating
NT2 selective breeding
NT1 animal welfare
NT1 feeding
NT2 fattening
NT3 extensive production
NT3 indoor fattening
NT3 intensive fattening
NT2 feed rations

LIVESTOCK SECTOR (C)
NT2 feed requirements
NT2 fodder conversion
NT1 herd improvement
NT1 livestock farming
NT2 extensive livestock farming
NT2 hand rearing
NT2 industrial rearing
NT1 livestock husbandry
NT2 controlled grazing
NT2 grazing
NT2 slurry removal
NT2 stabling
NT2 transhumance
NT1 milking

LIVESTOCK WASTES
BT1 agricultural wastes
BT2 pollution by agriculture
BT3 environmental policy
rt organic fertilizers

LIVING CONDITIONS
uf *social conditions*
NT1 living standards
NT2 basic needs
NT2 poverty
NT3 hunger
NT1 rural housing
NT2 farm dwellings
NT1 rural welfare
rt socioeconomic status

LIVING STANDARDS
uf *cost of living*
uf *standard of living*
BT1 living conditions
NT1 basic needs
NT1 poverty
NT2 hunger
rt income distribution

LOANS
BT1 credit transactions
NT1 international loans
NT1 private loans
NT2 bank loans
NT2 mortgages
NT1 public loans
rt credit
rt credit policy
rt outside capital
rt outside finance

LOCAL AUTHORITY AREAS
BT1 local government
BT2 administration
rt villages

LOCAL BYELAWS
BT1 legal sources

LOCAL GOVERNMENT
BT1 administration
NT1 local authority areas

LOCAL PLANNING
BT1 planning

LOCATION OF PRODUCTION
BT1 production policy
rt location theory

LOCATION THEORY
sf *theory*
BT1 economic theories
rt location of production

LONG TERM CREDIT
BT1 credit
BT2 credit transactions

LOSSES
BT1 farm results
rt debt

low calorie foods
USE **prepared foods**

low cost technology
USE **appropriate technology**

low income areas
USE **less favoured areas**

LUCERNE
BT1 feed crops
BT2 crops

LUXEMBOURG
sf benelux
BT1 western europe
BT2 europe
rt belgium
rt european communities

MACAO
uf macau
BT1 east asia
BT2 asia
rt portugal

macau
USE **macao**

machinery
USE **farm machinery**

MACHINERY CO-OPERATIVES
uf machinery pools
BT1 farm helper services
BT2 co-operative services

machinery pools
USE **machinery co-operatives**

MADAGASCAR
uf malagasy republic
BT1 southern africa
BT2 africa south of sahara
BT3 africa
rt french-speaking africa

MADEIRA
BT1 western europe
BT2 europe
rt portugal

MAIL ORDER SELLING
BT1 marketing techniques

mainland china
USE **china**

MAINTENANCE
BT1 mechanization

MAIZE
BT1 cereals
BT2 crops
NT1 grain maize
NT1 silo maize

MALADJUSTMENT
BT1 social adjustment
BT2 socialization
rt social crisis

malagasy republic
USE **madagascar**

MALAWI
BT1 east africa
BT2 africa south of sahara
BT3 africa
BT1 southern africa
rt commonwealth
rt english-speaking africa

MALAYSIA
BT1 south east asia
BT2 asia
rt commonwealth

MALDIVES
BT1 south east asia
BT2 asia

MALI
BT1 west africa
BT2 africa south of sahara
BT3 africa
rt french-speaking africa

MALTA
BT1 mediterranean countries
BT1 western europe
BT2 europe
rt commonwealth

MALTING BARLEY
BT1 barley
BT2 cereals
BT3 crops

MANAGEMENT
BT1 human resources
BT2 economic resources
BT3 economic situation
NT1 farm management
NT1 self management

MANAGEMENT BY OBJECTIVES
BT1 farm planning

MANAGEMENT GAMES
BT1 teaching methods

MANAGERS
BT1 occupations
rt farmers

MANDARINS
BT1 citrus fruits
BT2 fruits

MANGOES
BT1 tropical fruits
BT2 fruits

MANIOC
uf cassava
uf tapioca
BT1 root crops
BT2 crops

MANPOWER UNITS
BT1 agricultural manpower
rt economic units

MANUFACTURED FEEDS
BT1 feedingstuffs
NT1 concentrates
NT1 feed cake
NT1 feed-mixes
rt additives

MANURE
BT1 organic fertilizers
BT2 fertilizers

MAPS
uf atlases
BT1 publications
rt geography

MARES
BT1 horses

MARGARINE
BT1 simulated foods
rt butter
rt fat products

MARGINAL ANALYSIS
BT1 economic analysis
NT1 marginal costs
NT1 marginal returns

MARGINAL COSTS
BT1 marginal analysis
BT2 economic analysis

MARGINAL RETURNS
BT1 marginal analysis
BT2 economic analysis

margins
USE **marketing margins**

MARKET ECONOMICS
NT1 competition
NT2 imperfect competition
NT3 monopoly
NT3 monopsony
NT3 oligopoly
NT3 oligopsony
NT2 perfect competition
NT1 demand
NT1 market transparency
NT2 market intelligence
NT1 supply
NT1 supply and demand
rt economic theories
rt market economies

MARKET ECONOMIES
BT1 economic systems
NT1 competitive economies
rt market economics

MARKET INTELLIGENCE
BT1 market transparency
BT2 market economics

MARKET PLANNING
BT1 planning

MARKET PRICES
BT1 prices

MARKET REGULATIONS
BT1 marketing policy
BT2 agricultural policy
NT1 intervention

MARKET RESEARCH
NT1 consumer surveys
NT2 consumer motivation
NT2 consumer panels
NT1 market surveys
rt marketing
rt marketing channels

MARKET SEGMENTATION
BT1 marketing techniques

MARKET STABILIZATION
BT1 marketing policy
BT2 agricultural policy
rt buffer stocks

MARKET SURVEYS
BT1 market research

MARKET TRANSPARENCY
BT1 market economics
NT1 market intelligence

MARKETING
NT1 distribution
NT1 marketing channels
NT2 commodity exchanges
NT3 futures markets
NT2 merchants
NT3 comission agents
NT3 middlemen
NT2 retail marketing
NT3 direct marketing
NT3 street markets
NT3 supermarkets
NT2 wholesale marketing
NT3 buying groups
NT3 product markets
rt market research

MARKETING BOARDS
BT1 trade policy

MARKETING CHANNELS
uf distribution channels
BT1 marketing
NT1 commodity exchanges
NT2 futures markets
NT1 merchants

MARKETING CHANNELS (C)
NT2 comission agents
NT2 middlemen
NT1 retail marketing
NT2 direct marketing
NT2 street markets
NT2 supermarkets
NT1 wholesale marketing
NT2 buying groups
NT2 product markets
rt market research
rt marketing margins

marketing co-operatives
USE **co-operative marketing**

MARKETING MARGINS
uf margins
uf price spreads
rt distribution
rt marketing channels

MARKETING POLICY
uf common guarantee policy
BT1 agricultural policy
NT1 market regulations
NT2 intervention
NT1 market stabilization
NT1 price policy
NT2 compensatory amounts
rt funds
rt policy

MARKETING TECHNIQUES
NT1 auctions
NT1 bonuses
NT1 credit sales
NT1 discounts
NT1 door to door sales
NT1 mail order selling
NT1 market segmentation
NT1 sales promotion
NT2 labelling
NT3 quality labelling
NT4 standard labelling
NT3 trade marks
NT2 launching a product
NT2 product image
NT2 product presentation
NT2 publicity
NT3 advertizing
NT3 trade fairs
NT2 special offers
rt purchasing
rt sales

MARKETS
NT1 capital market
NT1 commodity markets
NT1 domestic markets
NT1 intra-community markets
NT1 labour market
NT2 labour mobility
NT1 land markets
NT2 land mobility
NT1 world markets

MARKOV CHAINS
BT1 non-optimizing methods
BT2 operations research

MARMALADES
BT1 fruit products

MARRIAGE
BT1 families
NT1 divorce
NT1 marriage age

MARRIAGE AGE
BT1 marriage
BT2 families

MARROWS
BT1 vegetables

MARTINIQUE
BT1 caribbean

MARTINIQUE (C)
BT2 america
rt france

MARXIST ECONOMICS
BT1 economic schools
BT2 economic theories

MASS MEDIA
NT1 films
NT1 newspapers
NT2 agricultural press
NT1 radio
NT1 television
rt communication
rt extension
rt information diffusion
rt leisure

MASS SOCIETY
BT1 societies

MATE
BT1 soft drinks
BT2 beverages

MATERNITY BENEFITS
BT1 social security
rt insurances

MATHEMATICS
NT1 boolean algebra
NT1 calculus
NT1 financial mathematics
NT2 actuarial mathematics
NT1 matrices
NT1 modern mathematics
rt cybernetics
rt systems analysis
rt theory of graphs

MATING
BT1 animal breeding
BT2 livestock sector

MATRIARCHY
BT1 family structure

MATRICES
BT1 mathematics

MAURITANIA
BT1 west africa
BT2 africa south of sahara
BT3 africa
rt french-speaking africa

MAURITIUS
BT1 southern africa
BT2 africa south of sahara
BT3 africa
rt commonwealth
rt english-speaking africa

MAXIMUM PRICES
BT1 prices

MEASUREMENT
BT1 evaluation

MEAT
NT1 beef
NT1 horse meat
NT1 lamb
NT1 mutton
NT1 offals
NT1 pigmeat
NT1 poultry meat
NT1 veal
rt food
rt meat and livestock industry
rt meat products

MEAT ANALOGUES
uf meat substitutes
uf textured vegetable proteins(tvp)
BT1 simulated foods

MEAT AND LIVESTOCK INDUSTRY
BT1 food industry
NT1 abattoirs
rt meat

MEAT PRODUCTS
NT1 bacon
NT1 dried meats
NT1 ham
NT1 sausages
rt food
rt meat
rt prepared foods

meat substitutes
USE **meat analogues**

mechanical handling
USE **handling**

MECHANIZATION
NT1 maintenance
rt agricultural engineering
rt automation
rt farm equipment

MEDICINAL HERBS
sf herbs
BT1 crops

mediterranean area
USE **mediterranean countries**

MEDITERRANEAN COUNTRIES
uf mediterranean area
uf southern europe
NT1 albania
NT1 algeria
NT1 cyprus
NT1 egypt
NT1 france
NT2 france (overseas depts and terr)
NT2 france (r. alsace)
NT2 france (r. aquitaine)
NT2 france (r. auvergne)
NT2 france (r. brittany)
NT2 france (r. burgundy)
NT2 france (r. centre)
NT2 france (r. champagne ardennes)
NT2 france (r. corsica)
NT2 france (r. franche comte)
NT2 france (r. ile de france)
NT2 france (r. languedoc roussillon)
NT2 france (r. limousin)
NT2 france (r. lorraine)
NT2 france (r. lower normandy)
NT2 france (r. midi pyrenees)
NT2 france (r. nord pas de calais)
NT2 france (r. pays de la loire)
NT2 france (r. picardy)
NT2 france (r. poitou charentes)
NT2 france (r. provence alpes cote d'az
NT2 france (r. rhone alpes)
NT2 france (r. upper normandy)
NT1 gibraltar
NT1 greece
NT1 israel
NT1 italy
NT2 italy (abruzzi)
NT2 italy (apulia)
NT2 italy (basilicata)
NT2 italy (calabria)
NT2 italy (campania)
NT2 italy (emilia-romagna)
NT2 italy (friuli-venezia giulia)
NT2 italy (latium)
NT2 italy (liguria)
NT2 italy (lombardy)
NT2 italy (marches)
NT2 italy (molise)
NT2 italy (piedmont)
NT2 italy (sardinia)
NT2 italy (sicily)
NT2 italy (trentino-alto adige)
NT2 italy (tuscany)
NT2 italy (umbria)
NT2 italy (valle d'aosta)
NT2 italy (veneto)

MEDITERRANEAN COUNTRIES (C)
NT1 lebanon
NT1 libya
NT1 malta
NT1 monaco
NT1 morocco
NT1 san marino
NT1 spain
NT1 syria
NT1 tunisia
NT1 turkey
NT1 yugoslavia

MEDIUM SIZED FARMS
BT1 types of farming
BT2 agricultural systems
rt family farms
rt farm size

MEDIUM TERM CREDIT
BT1 credit
BT2 credit transactions

melanesia
USE **oceania**

melioration
USE **land improvement**

MELONS
uf *cantaloupes*
BT1 fruits

MEMBERSHIP
BT1 groups
NT1 individuals and groups
rt organizations

MERCHANTS
BT1 marketing channels
BT2 marketing
NT1 comission agents
NT1 middlemen

METAYAGE
uf *share cropping*
BT1 tenure systems

METEOROLOGY
rt climate

METHODOLOGY

MEXICO
uf *united mexican states*
sf *latin america*
BT1 north america
BT2 america
rt lafta

MICROCLIMATE
BT1 climate

MICROECONOMIC ANALYSIS
NT1 factors of production
rt economic analysis

microfiches
USE **microform**

microfilms
USE **microform**

MICROFORM
uf *microfiches*
uf *microfilms*
BT1 information storage
BT2 information processing

MIDDLE CLASSES
BT1 social classes
BT2 social stratification
BT3 social structure

middle east
USE **west asia**

MIDDLEMEN
BT1 merchants
BT2 marketing channels
BT3 marketing

MIGRANT LABOUR
uf *foreign workers*
BT1 hired labour
BT2 agricultural manpower

MIGRATION
NT1 emigration
NT1 geographical mobility
NT2 commuting
NT2 rural-urban migration
NT2 seasonal migration
NT2 urban rural migration
NT1 immigration
NT1 rural depopulation

MILK
uf *cows milk*
NT1 skim milk
rt food
rt milk products

MILK FAT
uf *butterfat*
BT1 milk products

milk powder
USE **dried milk**

MILK PRODUCTS
NT1 butter
NT1 cheese
NT2 hard cheese
NT2 processed cheese
NT2 soft cheese
NT1 cream
NT1 cultured milks
NT2 buttermilk
NT2 yoghurt
NT1 dried milk
NT2 dried skim milk
NT1 evaporated milk
NT1 flavoured milks
NT1 milk fat
NT1 milk proteins
NT1 sweetened condensed milk
NT1 whey
rt food
rt milk

MILK PROTEINS
BT1 milk products

MILK SUBSTITUTES
uf *coffee whiteners*
uf *cream substitutes*
BT1 simulated foods

MILKING
BT1 livestock sector

MILLET
BT1 cereals
BT2 crops

MILLING INDUSTRY
BT1 food industry

MINERAL SUPPLEMENTS
BT1 additives

MINICOMPUTERS
BT1 computers
BT2 hardware

MINIMAL TILLAGE SYSTEMS
BT1 tillage
BT2 cultivation methods
rt no-tillage system

MINIMUM PRICES
BT1 prices

MINISTRIES OF AGRICULTURE
BT1 central government
BT2 administration

MIXED CROPPING
BT1 farm enterprises
rt diversification
rt mixed farming

MIXED ECONOMIES
BT1 economic systems
NT1 welfare economics

MIXED FARMING
BT1 types of farming
BT2 agricultural systems
rt mixed cropping

mixed integer programming
USE **integer programming**

mobility
SEE **labour mobility**
OR **land mobility**
OR **social mobility**

MODELS
NT1 dynamic models
NT1 econometric models
NT1 farm models
NT1 international trade models
NT1 transport models

MODERN MATHEMATICS
BT1 mathematics
rt theory of graphs

MODERNIZATION
BT1 social change
NT1 adoption of innovations
NT1 industrialization
NT1 technical progress
NT2 appropriate technology
NT2 technology transfers

MOLE-KILLER
BT1 pesticides

MONACO
BT1 mediterranean countries
BT1 western europe
BT2 europe

MONETARY AGREEMENTS
BT1 monetary policy
rt drawing rights

monetary compensatory amounts
USE **compensatory amounts**

MONETARY PARITY
uf *exchange rates*
BT1 monetary policy
NT1 devaluation
NT1 revalution

MONETARY POLICY
NT1 monetary agreements
NT1 monetary parity
NT2 devaluation
NT2 revalution
NT1 monetary reform
NT1 monetary unions
rt policy

MONETARY REFORM
BT1 monetary policy

MONETARY SITUATION
NT1 deflation
NT1 inflation

MONETARY SYSTEMS
NT1 drawing rights

MONETARY UNIONS
BT1 monetary policy
rt economic unions

MONGOLIA
uf *mongolia pr*
BT1 east asia
BT2 asia
rt cmea

mongolia pr
USE **mongolia**

MONOPOLY
BT1 imperfect competition
BT2 competition
BT3 market economics

MONOPSONY
BT1 imperfect competition
BT2 competition
BT3 market economics

MONTE CARLO METHOD
BT1 simulation
BT2 non-optimizing methods
BT3 operations research

MONTSERRAT
BT1 caribbean
BT2 america
rt united kingdom

MOROCCO
sf *africa (maghreb)*
BT1 arab countries
BT1 mediterranean countries
BT1 north africa
BT2 africa
rt french-speaking africa

MORTALITY
BT1 population change

MORTGAGES
BT1 private loans
BT2 loans
BT3 credit transactions
BT1 property transfers
BT2 land ownership

MOSHAVIM
BT1 collective farms
BT2 farms

MOTHERS
BT1 parents
BT2 family size
BT3 families

MOTIVATION
BT1 decision making

MOUNTAIN AREAS
uf *hill areas*
BT1 less favoured areas
BT2 regional agricultural policy
BT3 agricultural policy

MOZAMBIQUE
BT1 southern africa
BT2 africa south of sahara
BT3 africa
rt portugal

mucking out
USE **slurry removal**

MULTI-GENERATION FAMILIES
BT1 extended families
BT2 family structure

MULTI-PURPOSE CO-OPERATIVES
BT1 co-operatives

MULTICOLLINEARITY
BT1 statistical inference
BT2 statistical analysis
BT3 statistics

MULTINATIONAL CORPORATIONS
BT1 firms

multiple discriminant analysis
USE **discriminant analysis**

muscat and oman
USE **oman**

MUSTARD
BT1 oilseeds
BT2 crops

MUTTON
BT1 meat

mutual help
SEE **co-operation**
OR **farm helper services**

MUTUAL ORGANISATIONS
BT1 co-operatives
rt companies

NAMIBIA
uf *south west africa*
BT1 southern africa
BT2 africa south of sahara
BT3 africa
rt english-speaking africa

NANNY GOATS
BT1 goats

NATIONAL ACCOUNTING
BT1 aggregate accounts
rt accounting
rt gross national product
rt income

NATIONAL AGRICULTURAL POLICY
BT1 agricultural policy
rt policy

NATIONAL CHARACTER
BT1 nationalism

NATIONAL CONSCIOUSNESS
BT1 nationalism

NATIONAL CURRENCIES
BT1 currency

NATIONAL EXPENDITURE
BT1 aggregate data

NATIONAL FORESTS
BT1 forest ownership
BT1 public domain
BT2 land use systems
rt forest land

NATIONAL FUND
BT1 funds

national insurance
USE **social insurance**

NATIONAL PARKS
uf *national trust*
BT1 public domain
BT2 land use systems
rt open spaces

NATIONAL PLANNING
BT1 planning
rt nationalized enterprises

national trust
USE **national parks**

NATIONALISM
NT1 national character
NT1 national consciousness
rt nations
rt political systems

NATIONALIZED ENTERPRISES
BT1 firms
rt national planning

NATIONS
rt nationalism

NATURAL DISASTERS
BT1 environment
rt agricultural disasters

NATURAL FAUNA
BT1 wild life

NATURAL FLORA
BT1 wild life

NATURAL RESOURCES
BT1 economic resources
BT2 economic situation
rt land

NATURE CONSERVANCY
BT1 environmental policy

NAURU
BT1 oceania
rt commonwealth

near east
USE **west asia**

NECTARINES
BT1 peaches
BT2 stone fruits
BT3 fruits

NEIGHBOURHOOD
BT1 small groups
BT2 group size
BT3 groups

NEMATICIDES
BT1 pesticides

NEO-CLASSICAL ECONOMICS
BT1 economic schools
BT2 economic theories

NEPAL
BT1 south east asia
BT2 asia

NET RETURNS
BT1 farm results

net worth
USE **owners equity**

NETHERLANDS
uf *holland*
sf *benelux*
BT1 western europe
BT2 europe
NT1 netherlands (drenthe)
NT1 netherlands (friesland)
NT1 netherlands (gelderland)
NT1 netherlands (groningen)
NT1 netherlands (n.brabant)
NT1 netherlands (n.holland)
NT1 netherlands (overijssel)
NT1 netherlands (s.holland)
NT1 netherlands (s.limburg)
NT1 netherlands (utrecht)
NT1 netherlands (zeeland)
rt belgium
rt european communities
rt netherlands antilles
rt surinam

NETHERLANDS (DRENTHE)
BT1 netherlands
BT2 western europe
BT3 europe

NETHERLANDS (FRIESLAND)
BT1 netherlands
BT2 western europe
BT3 europe

NETHERLANDS (GELDERLAND)
BT1 netherlands

NETHERLANDS (GELDERLAND) (C)
BT2 western europe
BT3 europe

NETHERLANDS (GRONINGEN)
BT1 netherlands
BT2 western europe
BT3 europe

NETHERLANDS (N.BRABANT)
BT1 netherlands
BT2 western europe
BT3 europe

NETHERLANDS (N.HOLLAND)
BT1 netherlands
BT2 western europe
BT3 europe

NETHERLANDS (OVERIJSSEL)
BT1 netherlands
BT2 western europe
BT3 europe

NETHERLANDS (S.HOLLAND)
BT1 netherlands
BT2 western europe
BT3 europe

NETHERLANDS (S.LIMBURG)
BT1 netherlands
BT2 western europe
BT3 europe

NETHERLANDS (UTRECHT)
BT1 netherlands
BT2 western europe
BT3 europe

NETHERLANDS (ZEELAND)
BT1 netherlands
BT2 western europe
BT3 europe

NETHERLANDS ANTILLES
uf *antilles (netherlands)*
BT1 caribbean
BT2 america
rt netherlands

NETWORK ANALYSIS
BT1 plan implementation and evaluation
BT2 operations research

NEW CALEDONIA
BT1 pacific islands fr
BT2 oceania

NEW HEBRIDES
BT1 oceania
rt pacific islands fr
rt pacific islands gb

NEW ZEALAND
BT1 oceania
rt commonwealth
rt pacific islands nz

NEWSPAPERS
BT1 mass media
NT1 agricultural press

NICARAGUA
BT1 central america
BT2 america
rt cacm

NIGER
BT1 west africa
BT2 africa south of sahara
BT3 africa
rt french-speaking africa

NIGERIA
BT1 west africa
BT2 africa south of sahara
BT3 africa
rt commonwealth

NIGERIA (C)
rt english-speaking africa

NITROGEN
BT1 inorganic fertilizers
BT2 fertilizers
NT1 urea

NO-TILLAGE SYSTEM
BT1 tillage
BT2 cultivation methods
rt minimal tillage systems

NOMADS
uf *pastoral nomads*
BT1 pastoral society
BT2 societies
rt deserts
rt transhumance

non-convertible currencies
USE **blocked currencies**

NON-FARM INCOME
BT1 farmers income
BT2 farm results
rt off-farm employment
rt peasant workers

NON-FOOD INDUSTRIES
NT1 leather industry
NT1 paper industry
NT1 rubber industry
NT1 textile industries
NT2 cotton industry
NT3 cotton ginning
NT2 wool industry
NT1 tobacco industry
NT1 wood industry
rt industry
rt processing

NON-GAINFULLY EMPLOYED
BT1 population structure
BT2 population
NT1 retired people
NT1 unemployed

non-governmental organizations
USE **private organizations**

NON-LINEAR PROGRAMMING
BT1 programming

NON-MARKETING PREMIUMS
BT1 subsidies
BT2 agricultural financial policy

NON-OPTIMIZING METHODS
BT1 operations research
NT1 cost benefit analysis
NT1 farm budgeting
NT1 game theory
NT1 input-output analysis
NT1 markov chains
NT1 ppbs
NT1 simulation
NT2 monte carlo method
NT1 systems analysis
rt planning
rt programming

NON-PARAMETRIC METHODS
BT1 econometrics

NON-TARIFF BARRIERS TO TRADE
NT1 tax agreements
rt trade barriers

NON-TRANSFERABLE PROPERTY
rt indivisible estates
rt land mobility

NORM PRICES
(cap price for tobacco)
BT1 prices

NORMATIVE PLANNING
BT1 planning

NORTH AFRICA
BT1 africa
NT1 algeria
NT1 egypt
NT1 libya
NT1 morocco
NT1 tunisia
rt arab countries

NORTH AMERICA
BT1 america
NT1 bermuda
NT1 canada
NT1 greenland
NT1 mexico
NT1 st pierre and miquelon
NT1 usa

north korea
USE **korea dpr**

north vietnam
USE **vietnam**

northern europe
USE **scandinavia**

NORWAY
BT1 scandinavia
BT2 western europe
BT3 europe
rt antarctica
rt efta

NOXIOUS SUBSTANCES
uf *chemical residues*
rt pollution by agriculture
rt pollution of agriculture

NUCLEAR FAMILIES
BT1 family structure
NT1 one generation families
NT1 two generation families

NURSE COWS
BT1 cows
BT2 cattle

nursery plants
USE **plants**

NUT PRODUCTS
BT1 fruit products

NUTRITION
rt eating habits
rt food policy
rt food supply
rt health
rt proteins

NUTS
NT1 almonds
NT2 bitter almonds
NT2 sweet almonds
NT1 brazil nuts
NT1 cashew nuts
NT1 chestnuts
NT1 coconuts
NT1 cola nuts
NT1 hazelnuts
NT1 hickory nuts
NT1 pecans
NT1 pistachios
NT1 shea nuts
NT1 walnuts
rt fruits

OATMEAL
BT1 flour
BT2 cereal products

OATS
BT1 cereals
BT2 crops

OATS (C)
rt porridge oats

OCCUPATIONAL CHANGE
BT1 employment
rt retraining
rt rural depopulation

OCCUPATIONAL PRESTIGE
BT1 status
BT2 social stratification
BT3 social structure

OCCUPATIONAL STATUS
BT1 status
BT2 social stratification
BT3 social structure

OCCUPATIONS
uf professions
NT1 farm workers
NT1 farmers
NT1 managers
NT1 teachers
NT1 workers

OCEANIA
uf australasia
uf melanesia
sf pacific islands
sf polynesia
NT1 australia
NT1 fiji
NT1 nauru
NT1 new hebrides
NT1 new zealand
NT1 pacific islands au
NT1 pacific islands fr
NT2 french polynesia
NT2 new caledonia
NT1 pacific islands gb
NT1 pacific islands nz
NT1 pacific islands us
NT1 papua new guinea
NT1 tonga
NT1 western samoa

OECD
(organization for economic co-operation and development)
BT1 organizations

OFF-FARM EMPLOYMENT
BT1 employment
rt non-farm income
rt part-time farming
rt peasant workers

OFFALS
BT1 meat

oils
USE **oils and fats**

OILS AND FATS
uf fats
uf oils
NT1 animal fats
NT1 fat products
NT1 vegetable oils
NT2 coconut oils
NT2 corn oils
NT2 cottonseed oils
NT2 groundnut oils
NT2 olive oil
NT2 palm oil
NT2 soy oils
NT2 sunflower seed oils
rt oils and fats industry

OILS AND FATS INDUSTRY
BT1 food industry
rt oils and fats

OILSEEDS
BT1 crops
NT1 castor beans
NT1 groundnuts

OILSEEDS (C)
NT1 linseed
NT1 mustard
NT1 rape
NT1 sesame
NT1 soybeans
rt industrial crops

OKRA
BT1 vegetables

OLD AGE
BT1 age

OLD AGE BENEFITS
uf pensions
BT1 social security
rt insurances

OLIGOPOLY
BT1 imperfect competition
BT2 competition
BT3 market economics

OLIGOPSONY
BT1 imperfect competition
BT2 competition
BT3 market economics

OLIVE OIL
BT1 vegetable oils
BT2 oils and fats

OLIVES
BT1 stone fruits
BT2 fruits

OMAN
uf muscat and oman
BT1 west asia
BT2 asia

ON FARM CONSUMPTION
BT1 consumption
rt eating habits

ON FARM TRAINING
BT1 vocational training
BT2 education

ON LINE
BT1 data processing

ONE GENERATION FAMILIES
BT1 nuclear families
BT2 family structure

ONIONS
BT1 vegetables

OPEN SPACES
uf green belt
BT1 physical planning
BT2 planning
rt landscape
rt national parks

OPERATIONS RESEARCH
sf analysis
NT1 bayesian theory
NT1 non-optimizing methods
NT2 cost benefit analysis
NT2 farm budgeting
NT2 game theory
NT2 input-output analysis
NT2 markov chains
NT2 ppbs
NT2 simulation
NT3 monte carlo method
NT2 systems analysis
NT1 optimization methods
NT2 international trade models
NT2 transport models
NT1 plan implementation and evaluation
NT2 cash flow analysis
NT2 critical path analysis
NT2 network analysis
NT2 pert

OPERATIONS RESEARCH (C)
NT2 queueing theory
NT1 project control
NT2 altai programme
NT2 budgetary control
rt cybernetics
rt econometrics
rt planning
rt sociological method
rt theory of graphs

OPINIONS
BT1 behaviour

OPTIMIZATION METHODS
BT1 operations research
NT1 international trade models
NT1 transport models
rt integer programming
rt planning
rt programming

OPTION TO PURCHASE
rt contracts
rt termination of leases

orange juice
USE **fruit juices**

ORANGES
BT1 citrus fruits
BT2 fruits

ORCHARDS
BT1 land
rt fruit growing

ORGANIC FARMING
BT1 cropping systems

ORGANIC FERTILIZERS
BT1 fertilizers
NT1 green manure
NT1 guano
NT1 manure
NT1 slurry
rt livestock wastes

ORGANIZATION OF RESEARCH
uf research organization
NT1 research institutions
NT2 experimental stations
NT2 laboratories
NT2 pilot farms
NT2 research associations
NT2 research institutes
NT1 research personnel
NT2 research teams
NT2 research workers
NT3 agricultural economists
NT3 rural sociologists

ORGANIZATION OF WORK
BT1 labour economics
NT1 labour allocation

ORGANIZATIONS
sf associations
sf institutions
NT1 acp
NT1 cacm
NT1 cmea
NT1 efta
NT1 european communities
NT1 european parliament
NT1 fao
NT1 lafta
NT1 oecd
NT1 un
NT1 unesco
NT1 who
NT1 world bank
rt interest groups
rt membership

ORNAMENTAL PLANTS
BT1 crops
NT1 flowers

ORNAMENTAL PLANTS (C)
NT2 bulbs

OUTSIDE CAPITAL
BT1 liabilities
BT2 balance sheets
BT3 balance sheet analysis
rt loans

OUTSIDE FINANCE
BT1 finance
rt credit
rt debt
rt loans

overproduction
USE **surpluses**

OVICIDES
BT1 pesticides

OWNER OCCUPANCY
BT1 tenure systems

OWNERS EQUITY
uf *net worth*
BT1 liabilities
BT2 balance sheets
BT3 balance sheet analysis

OWNERSHIP
uf *property*
uf *real estate*
NT1 co-ownership
NT1 ownership limitation
NT2 adverse possession
NT2 compulsory purchase
NT2 easements
NT1 private ownership
NT1 public ownership
NT1 usufruct
rt land consolidation
rt land ownership

OWNERSHIP CONDITIONS
BT1 social stratification
BT2 social structure
rt land ownership

OWNERSHIP LIMITATION
BT1 ownership
NT1 adverse possession
NT1 compulsory purchase
NT1 easements

pacific islands
SEE **oceania**
OR **pacific islands au**
OR **pacific islands fr**
OR **pacific islands gb**
OR **pacific islands nz**
OR **pacific islands us**

PACIFIC ISLANDS AU
sf *pacific islands*
BT1 oceania
rt australia

PACIFIC ISLANDS FR
sf *pacific islands*
BT1 oceania
NT1 french polynesia
NT1 new caledonia
rt france
rt new hebrides

PACIFIC ISLANDS GB
sf *pacific islands*
BT1 oceania
rt new hebrides
rt united kingdom

PACIFIC ISLANDS NZ
sf *pacific islands*
BT1 oceania
rt new zealand

PACIFIC ISLANDS US
sf *pacific islands*
BT1 oceania
rt usa

PACKAGING
uf *packing*
NT1 bottling
NT1 canning
rt labelling
rt packs
rt preservation

packing
USE **packaging**

PACKS
uf *bales*
uf *cartons*
uf *jars*
uf *sacks*
rt packaging
rt product presentation

paddock grazing
USE **controlled grazing**

PAKISTAN
BT1 south east asia
BT2 asia

PALM OIL
BT1 vegetable oils
BT2 oils and fats

pampas
USE **prairies**

PANAMA
BT1 central america
BT2 america

PANAMA CANAL ZONE
BT1 central america
BT2 america

PAPER INDUSTRY
BT1 non-food industries
rt forest products

paprika
USE **peppers**

PAPUA NEW GUINEA
BT1 oceania
rt commonwealth

PARAGUAY
BT1 south america
BT2 america
rt lafta

PARAMETRIC METHODS
BT1 econometrics
NT1 simultaneous equation analysis
NT1 single equation analysis
NT2 consumption functions
NT2 demand functions
NT2 elasticities
NT2 flexibility
NT2 investment functions
NT2 production functions
NT2 savings functions
NT2 supply functions
NT2 utility functions

PARAMETRIC PROGRAMMING
BT1 programming

PARASITES
rt animal diseases
rt pesticides
rt plant diseases

parastatal organizations
USE **semi-private organizations**

PARCEL SIZE
BT1 farm size
rt land consolidation

PARENT CHILD RELATIONSHIPS
BT1 generation conflict
BT2 family life

PARENTS
BT1 family size
BT2 families
NT1 fathers
NT1 mothers

PARITY
BT1 agricultural situation
rt income distribution

PARLIAMENT
BT1 democracy
BT2 political systems
rt european parliament

PARSNIPS
BT1 vegetables

PART TIME LABOUR
BT1 hired labour
BT2 agricultural manpower

PART-TIME FARMING
BT1 types of farming
BT2 agricultural systems
NT1 peasant workers
rt ancillary enterprises
rt off-farm employment

PARTITION OF ESTATES
BT1 succession
BT2 acquisition of ownership
rt indemnification

PARTNERSHIPS
BT1 family structure

PARTRIDGES
BT1 game birds
BT2 poultry

PASSION FRUIT
BT1 tropical fruits
BT2 fruits

PASTA
BT1 cereal products

PASTA INDUSTRY
BT1 bakery industry
BT2 food industry
rt starch industry

pastoral nomads
USE **nomads**

PASTORAL SOCIETY
BT1 societies
NT1 nomads

PASTURES
uf *grasslands*
BT1 land
NT1 permanent pastures
NT1 prairies
NT1 temporary grasslands

PATRIARCHY
BT1 family structure

PAWPAWS
BT1 tropical fruits
BT2 fruits

payment by results
USE **piece work wages**

payment schemes
SEE **remuneration**
OR **wages**

PEACHES
BT1 stone fruits
BT2 fruits
NT1 nectarines

peanuts
USE **groundnuts**

PEARL BARLEY
BT1 cereal products

PEARS
BT1 pome fruits
BT2 fruits

PEAS
BT1 legumes
BT2 vegetables
NT1 forage peas

PEASANT WORKERS
BT1 part-time farming
BT2 types of farming
BT3 agricultural systems
rt non-farm income
rt off-farm employment

PEASANTRY
BT1 agricultural society
BT2 societies

PECANS
BT1 nuts

pensions
USE **old age benefits**

PEPPERS
uf capsicums
uf paprika
BT1 vegetables

PERENNIAL CROPPING
BT1 cropping systems

PERFECT COMPETITION
BT1 competition
BT2 market economics

PERIOD OF TENANCY
rt agricultural tenure

PERIODICALS
BT1 publications
NT1 secondary journals
NT2 bibliographies

PERMANENT LABOUR
BT1 hired labour
BT2 agricultural manpower

PERMANENT PASTURES
BT1 pastures
BT2 land

PERSIAN GULF STATES
BT1 asia
NT1 bahrain
NT1 kuwait
NT1 qatar
NT1 united arab emirates

PERSIMMONS
BT1 fruits

PERT
BT1 plan implementation and evaluation
BT2 operations research

PERU
BT1 south america
BT2 america
rt lafta

PEST CONTROL
BT1 plant husbandry
BT2 cultivation methods

PESTICIDES
NT1 acaricides
NT1 bactericides
NT1 corvicides
NT1 fungicides
NT1 herbicides
NT1 insecticides
NT1 larvicides
NT1 mole-killer
NT1 nematicides
NT1 ovicides
NT1 rodenticides
NT1 selective herbicides
NT1 virus killers
rt farm inputs
rt parasites
rt phytotoxicity

PHEASANTS
BT1 game birds
BT2 poultry

PHILIPPINES
BT1 south east asia
BT2 asia

PHOSPHATES
BT1 inorganic fertilizers
BT2 fertilizers

PHYSICAL PLANNING
BT1 planning
NT1 open spaces
NT1 zoning
rt environmental policy
rt structural policy

phytopathology
USE **plant diseases**

PHYTOTOXICITY
rt pesticides

PIECE WORK WAGES
uf payment by results
BT1 remuneration
BT2 labour economics

PIG FARMING
BT1 livestock enterprises
BT2 farm enterprises
rt pigs

PIGEONS
BT1 game birds
BT2 poultry

PIGLETS
uf weaner pigs
BT1 pigs

PIGMEAT
uf pork
BT1 meat
rt bacon
rt ham

PIGS
NT1 boars
NT1 piglets
NT1 sows
rt livestock
rt pig farming

PILOT FARMS
BT1 research institutions
BT2 organization of research

pineapple juice
USE **fruit juices**

PINEAPPLES
BT1 tropical fruits
BT2 fruits

PISTACHIOS
BT1 nuts

PLAN COMPTABLE
(in france only)
BT1 book-keeping

PLAN IMPLEMENTATION AND EVALUATION
BT1 operations research
NT1 cash flow analysis
NT1 critical path analysis
NT1 network analysis
NT1 pert
NT1 queueing theory
rt planning

PLANNING
NT1 agricultural planning
NT1 development plans
NT2 development projects
NT1 indicative planning
NT1 interregional planning
NT1 intersectoral planning
NT1 local planning
NT1 market planning
NT1 national planning
NT1 normative planning
NT1 physical planning
NT2 open spaces
NT2 zoning
NT1 regional planning
NT1 rural planning
NT2 building controls
NT2 central places
NT1 sectoral planning
rt evaluation
rt farm planning
rt non-optimizing methods
rt operations research
rt optimization methods
rt plan implementation and evaluation
rt work planning

PLANNING OF RESEARCH
BT1 research policy
NT1 research projects

PLANT BREEDING
sf breeding
NT1 hybridization
NT1 plants
NT1 seeds
NT1 varieties
NT2 high yielding varieties

PLANT DISEASES
uf phytopathology
sf diseases
rt parasites
rt public health legislation

PLANT HUSBANDRY
BT1 cultivation methods
NT1 biological control
NT1 grafting
NT1 integrated plant protection
NT1 pest control
NT1 pruning
NT1 thinning
NT1 weed control

PLANT PROTECTION
NT1 burning
NT1 fumigation
NT1 spraying
NT1 weeding
rt integrated plant protection

PLANTATIONS
BT1 types of farming
BT2 agricultural systems

PLANTING
BT1 tillage
BT2 cultivation methods
rt plants
rt replanting

PLANTS
uf nursery plants
BT1 plant breeding

PLANTS (C)
rt farm inputs
rt planting

ploughback
USE **self finance**

PLOUGHING
BT1 tillage
BT2 cultivation methods

PLUMS
BT1 stone fruits
BT2 fruits

PLURAL SOCIETIES
BT1 societies

POLAND
BT1 eastern europe
BT2 europe
rt cmea

POLDER
BT1 reclaimed land
BT2 land

POLICY
(use only for policies not specifically listed)
rt agricultural financial policy
rt agricultural policy
rt cap
rt credit policy
rt development policy
rt educational policy
rt environmental policy
rt food policy
rt forest policy
rt intervention
rt investment policy
rt marketing policy
rt monetary policy
rt national agricultural policy
rt price policy
rt production policy
rt regional agricultural policy
rt research policy
rt rural social policy
rt social policy
rt structural policy
rt support measures
rt trade policy

POLITICAL ATTITUDES
BT1 attitudes
BT2 behaviour
rt collective behaviour

POLITICAL POWER
BT1 power
BT2 social systems

POLITICAL SYSTEMS
NT1 capitalism
NT1 communism
NT1 democracy
NT2 parliament
NT1 dictatorship
NT1 federalism
NT1 feudalism
NT1 imperialism
NT2 colonialism
NT1 socialism
rt nationalism

POLLUTION BY AGRICULTURE
BT1 environmental policy
NT1 agricultural wastes
NT2 crop residues
NT2 livestock wastes
NT1 waste disposal
rt environment
rt noxious substances
rt slurry removal

POLLUTION CONTROL
BT1 environmental policy

POLLUTION OF AGRICULTURE
BT1 environmental policy
NT1 air pollution
NT1 water pollution
rt environment
rt noxious substances

polynesia
SEE **french polynesia**
OR **oceania**

POME FRUITS
BT1 fruits
NT1 apples
NT1 pears
NT1 quinces

POMEGRANATES
BT1 fruits
rt commonwealth

POPULATION
NT1 population structure
NT2 agricultural population
NT2 non-gainfully employed
NT3 retired people
NT3 unemployed
NT2 rural population
NT2 urban population
NT2 working population
NT3 agricultural work force
rt demography

POPULATION CENSUSES
BT1 censuses

POPULATION CHANGE
NT1 birth rate
NT1 family planning
NT1 health
NT1 mortality
NT1 population decrease
NT1 population density
NT1 population equilibrium
NT1 population forecasts
NT1 population growth
NT1 population pressure
rt change

population control
USE **family planning**

POPULATION DECREASE
BT1 population change

POPULATION DENSITY
BT1 population change

POPULATION EQUILIBRIUM
BT1 population change

POPULATION FORECASTS
BT1 population change

POPULATION GROWTH
BT1 population change

POPULATION PRESSURE
BT1 population change

POPULATION PYRAMIDS
BT1 age structure
BT2 age

POPULATION STRUCTURE
BT1 population
NT1 agricultural population
NT1 non-gainfully employed
NT2 retired people
NT2 unemployed
NT1 rural population
NT1 urban population
NT1 working population
NT2 agricultural work force
rt structure

pork
USE **pigmeat**

PORRIDGE OATS
BT1 cereal products
rt oats

porto rico
USE **puerto rico**

PORTUGAL
BT1 western europe
BT2 europe
rt angola
rt azores
rt cape verde
rt efta
rt macao
rt madeira
rt mozambique
rt portuguese timor
rt sao tome and principe

PORTUGUESE TIMOR
BT1 south east asia
BT2 asia
rt portugal

POSSESSION
BT1 acquisition of ownership

POST-AGRICULTURAL SECTOR
BT1 agro-industrial sector
rt food industry

POTASH
BT1 inorganic fertilizers
BT2 fertilizers

POTATO PRODUCTS
BT1 vegetable products

POTATOES
BT1 root crops
BT2 crops
NT1 feed potatoes
NT1 industrial potatoes
NT1 ware potatoes

POULTRY
NT1 broilers
NT2 cockerels
NT2 day old chicks
NT2 hens
NT1 ducks
NT1 game birds
NT2 partridges
NT2 pheasants
NT2 pigeons
NT1 geese
NT1 turkeys
rt livestock
rt poultry farming

POULTRY FARMING
BT1 livestock enterprises
BT2 farm enterprises
rt poultry

POULTRY MEAT
BT1 meat

POVERTY
BT1 living standards
BT2 living conditions
NT1 hunger

POWER
BT1 social systems
NT1 authority
NT1 leadership
NT1 political power
rt pressure groups

PPBS
BT1 non-optimizing methods
BT2 operations research

PRACTICAL EDUCATION
BT1 vocational training
BT2 education

PRACTICE
BT1 case law
rt ec legal decisions

PRAIRIES
uf *pampas*
uf *savannah*
uf *steppe*
BT1 pastures
BT2 land

PRE-AGRICULTURAL SECTOR
BT1 agro-industrial sector
rt input industries

PRECIPITATION
BT1 climatic factors
NT1 hail
NT1 rainfall
NT1 snow

PREFERENTIAL INTEREST RATES
BT1 interest rates
BT2 credit control
BT3 credit policy
rt public loans

PREFERENTIAL TARIFFS
BT1 tariffs

PREPARED FOODS
(foods which are more or less ready to consume, with or without heating or addition of liquids)
uf *canned meals*
uf *convenience foods*
uf *high calory foods*
uf *low calorie foods*
uf *slimming foods*
NT1 dietetic foods
NT2 infant foods
NT1 dried foods
NT2 freeze-dried foods
NT1 instant foods
NT2 instant cocoa
NT2 instant coffee
NT2 instant tea
rt fish products
rt food
rt fruit products
rt meat products
rt vegetable products

PRESERVATION
NT1 cooling
NT1 drying
NT1 freezing
rt additives
rt packaging
rt processing
rt silage

PRESERVATIVES
BT1 additives

PRESSURE GROUPS
BT1 interest groups
BT2 groups
rt power

PREVENTION OF FRAUD
BT1 consumer protection

PRICE CONTROLS
BT1 price formation

PRICE DISCRIMINATION
BT1 discrimination
BT2 protection

PRICE ELASTICITY
BT1 price formation
rt elasticities

PRICE FIXING
BT1 price formation

PRICE FORMATION
NT1 price controls

PRICE FORMATION (C)
NT1 price elasticity
NT1 price fixing
NT1 price stabilization
NT1 price support

PRICE POLICY
BT1 marketing policy
BT2 agricultural policy
NT1 compensatory amounts
rt agricultural prices
rt policy
rt prices
rt target prices

price spreads
USE **marketing margins**

PRICE STABILIZATION
BT1 price formation

PRICE SUPPORT
BT1 price formation

PRICES
NT1 activating prices
NT1 agricultural prices
NT2 guaranteed minimum producer prices
NT2 guide prices
NT2 intervention prices
NT1 average prices
NT1 basic prices
NT1 buying prices
NT1 consumer prices
NT1 controlled prices
NT1 cost prices
NT1 food prices
NT1 frozen prices
NT1 guide prices
NT1 input prices
NT1 market prices
NT1 maximum prices
NT1 minimum prices
NT1 norm prices
NT1 producer prices
NT1 reference prices
NT1 retail prices
NT1 selling prices
NT1 sluicegate prices
NT1 target prices
NT1 threshold prices
NT1 wholesale prices
NT1 withdrawal prices
rt price policy

PRIMARY EDUCATION
BT1 schools
BT2 education

primary literature
USE **publications**

PRIMARY PRODUCTS
uf *raw materials*
BT1 products
NT1 forest products

PRIMARY SECTOR
sf *sectors of production*
BT1 economic sectors
NT1 farm sector

PRIMITIVE ROTATION FARMING
BT1 cropping systems

principe
USE **sao tome and principe**

PRIVATE COMPANIES
BT1 private firms
BT2 firms

PRIVATE FARMS
BT1 farms

PRIVATE FIRMS
BT1 firms
NT1 companies

PRIVATE FIRMS (C)
NT1 private companies
rt co-operatives

PRIVATE FORESTS
BT1 forest ownership
rt forest land

PRIVATE INVESTMENT
BT1 investment
BT2 agricultural situation

PRIVATE LOANS
BT1 loans
BT2 credit transactions
NT1 bank loans
NT1 mortgages

PRIVATE ORGANIZATIONS
uf *non-governmental organizations*
BT1 administration

PRIVATE OWNERSHIP
BT1 ownership

PRIVATE RESEARCH
BT1 research

PROBABILITY ANALYSIS
BT1 stochastic processes
BT2 statistics

problem areas
USE **less favoured areas**

PROCESSED CHEESE
BT1 cheese
BT2 milk products

PROCESSED PRODUCTS
BT1 products
rt processing

PROCESSING
rt food industry
rt input industries
rt non-food industries
rt preservation
rt processed products

PRODUCER CO-OPERATIVES
BT1 co-operatives
rt collective farms

PRODUCER GROUPS
BT1 co-operative farm enterprises

PRODUCER PRICES
BT1 prices
rt agricultural prices

PRODUCT IMAGE
BT1 sales promotion
BT2 marketing techniques

PRODUCT MARKETS
BT1 wholesale marketing
BT2 marketing channels
BT3 marketing

PRODUCT PRESENTATION
BT1 sales promotion
BT2 marketing techniques
rt packs

PRODUCTION
NT1 domestic production
rt production policy
rt supply

PRODUCTION ACCOUNTS
BT1 economic accounts

PRODUCTION CONTROLS
BT1 production policy
NT1 agricultural guidelines
NT1 agro-industrial relations
NT1 surpluses

PRODUCTION CONTROLS (C)
rt quotas

PRODUCTION COSTS
BT1 costs
BT2 cost analysis
NT1 actual costs
NT1 estimated costs
NT1 fixed costs
NT1 special costs
NT1 total costs
NT1 variable costs
rt cost prices

PRODUCTION FUNCTIONS
BT1 single equation analysis
BT2 parametric methods
BT3 econometrics

production materials
USE **farm inputs**

PRODUCTION POLICY
NT1 adjustment of production
NT2 agricultural adjustment
NT2 dairy-beef conversion schemes
NT1 land diversion
NT1 location of production
NT1 production controls
NT2 agricultural guidelines
NT2 agro-industrial relations
NT2 surpluses
rt agricultural policy
rt policy
rt production
rt structural policy

PRODUCTION POSSIBILITIES
BT1 economic resources
BT2 economic situation
rt utilization of capacity

PRODUCTION STRUCTURE
BT1 agricultural structure
rt structure

PRODUCTION UNIT
BT1 economic units

PRODUCTIVITY
BT1 agricultural situation
BT1 profit and loss analysis
NT1 capital productivity
NT1 labour productivity
NT1 land productivity
rt yields

PRODUCTS
NT1 agricultural products
NT1 by-products
NT1 commodities
NT1 food
NT1 primary products
NT2 forest products
NT1 processed products

PROFESSIONAL CODES
BT1 legal sources

PROFESSIONAL SERVICES
BT1 tertiary sector
BT2 economic sectors
NT1 veterinary services
rt advisory centres
rt agricultural engineering

professions
USE **occupations**

PROFIT AND LOSS ANALYSIS
NT1 liquidity
NT1 productivity
NT2 capital productivity
NT2 labour productivity
NT2 land productivity
NT1 profitability
NT2 break-even point
NT2 gross margins

PROFIT AND LOSS ANALYSIS (C)
rt efficiency

PROFITABILITY
sf results
BT1 profit and loss analysis
NT1 break-even point
NT1 gross margins

PROGRAMMED LEARNING
BT1 teaching methods

PROGRAMMING
NT1 computer programming
NT1 concave programming
NT1 convex programming
NT1 dynamic programming
NT1 integer programming
NT1 linear programming
NT1 non-linear programming
NT1 parametric programming
NT1 quadratic programming
NT1 recursive programming
NT1 stochastic programming
rt non-optimizing methods
rt optimization methods
rt project control

PROJECT CONTROL
BT1 operations research
NT1 altai programme
NT1 budgetary control
rt programming

PROJECTIONS
BT1 forecasts

proletariat
USE **working classes**

property
USE **ownership**

PROPERTY PROTECTION
BT1 acquisition of ownership

property transfer tax
USE **stamp duty**

PROPERTY TRANSFERS
BT1 land ownership
NT1 inheritance
NT1 land transfers
NT1 mortgages

PROTECTION
NT1 discrimination
NT2 boycotts
NT2 dumping
NT2 embargos
NT2 price discrimination
NT1 quotas
NT1 trade barriers
NT2 export controls

protein feeds
USE **concentrates**

PROTEINS
NT1 animal proteins
NT1 synthetic proteins
NT1 vegetable proteins
rt nutrition

PRUNES
BT1 dried fruit
BT2 fruit products

PRUNING
BT1 plant husbandry
BT2 cultivation methods

PUBLIC AUTHORITIES
BT1 administration

PUBLIC DEBT
BT1 debt
BT2 credit transactions

PUBLIC DEBT (C)
rt public finance

PUBLIC DOMAIN
BT1 land use systems
NT1 common lands
NT1 national forests
NT1 national parks

PUBLIC ENTERPRISES
BT1 firms

PUBLIC FINANCE
NT1 agricultural budget
NT1 fiscal policy
NT2 direct taxation
NT3 capital taxation
NT4 land tax
NT3 company tax
NT3 income tax
NT3 inheritance tax
NT2 indirect taxation
NT3 exise duties
NT3 stamp duty
NT3 value added tax
NT2 taxes
rt public debt
rt public loans

PUBLIC FORESTS
BT1 forest ownership
rt forest land

PUBLIC HEALTH LEGISLATION
NT1 food inspection
rt animal diseases
rt plant diseases

PUBLIC INVESTMENT
BT1 investment
BT2 agricultural situation

public lendings
USE **public loans**

PUBLIC LOANS
uf public lendings
BT1 loans
BT2 credit transactions
rt preferential interest rates
rt public finance

PUBLIC OWNERSHIP
uf state ownership
BT1 ownership

PUBLIC RELATIONS
BT1 information diffusion

PUBLIC SERVICES
BT1 tertiary sector
BT2 economic sectors

PUBLICATIONS
uf primary literature
NT1 books
NT1 dictionary
NT1 directories
NT1 grey literature
NT1 maps
NT1 periodicals
NT2 secondary journals
NT3 bibliographies
NT1 reports
NT1 thesis
rt congresses

PUBLICITY
BT1 sales promotion
BT2 marketing techniques
NT1 advertizing
NT1 trade fairs

PUERTO RICO
uf porto rico
BT1 caribbean
BT2 america
BT1 central america

PUERTO RICO (C)
rt usa

PUMPKINS
BT1 vegetables

pupils
USE **students**

PURCHASING
BT1 trade
rt marketing techniques
rt turnover

PURCHASING HABITS
BT1 consumer behaviour
BT2 economic sociology

QATAR
BT1 persian gulf states
BT2 asia

QUADRATIC PROGRAMMING
BT1 programming

QUALITY
uf sensory evaluation+
rt quality labelling
rt quality standards

QUALITY CONTROLS
BT1 consumer protection
NT1 designation of origin
NT1 labelling controls
rt food inspection

QUALITY LABELLING
BT1 labelling
BT2 sales promotion
BT3 marketing techniques
NT1 standard labelling
rt quality
rt trade marks

QUALITY STANDARDS
BT1 standards
rt quality

QUALITY WINE
BT1 wine
BT2 alcoholic beverages
BT3 beverages

QUANTITY CONTROLS
BT1 standards

QUEUEING THEORY
uf waiting line
sf theory
BT1 plan implementation and evaluation
BT2 operations research

QUINCES
BT1 pome fruits
BT2 fruits

QUOTAS
BT1 protection
rt production controls
rt supply
rt surpluses

RABBITS
rt small animal rearing

RADIO
BT1 mass media
rt audiovisual aids

RADISHES
BT1 vegetables

RAIL TRANSPORT
BT1 transport

RAINFALL
BT1 precipitation
BT2 climatic factors

RAISINS
BT1 dried fruit
BT2 fruit products

RAMS
BT1 sheep

ranching
USE **extensive livestock farming**

RAPE
BT1 oilseeds
BT2 crops

RASPBERRIES
BT1 berries
BT2 fruits

RATIONALIZATION
BT1 socioeconomic organization

raw materials
USE **primary products**

re-afforestation
USE **afforestation**

real estate
USE **ownership**

recession
USE **economic depression**

RECLAIMED LAND
BT1 land
NT1 polder
rt accession

recreation
USE **leisure**

RECURSIVE PROGRAMMING
BT1 programming

RED CABBAGES
BT1 cabbages
BT2 vegetables

REDCURRANTS
BT1 berries
BT2 fruits

REFERENCE PRICES
BT1 prices

REGIONAL ACCOUNTING
BT1 aggregate accounts
rt accounting

REGIONAL AGRICULTURAL POLICY
BT1 agricultural policy
NT1 less favoured areas
NT2 mountain areas
rt funds
rt policy

REGIONAL DEVELOPMENT
BT1 economic development
rt funds

REGIONAL GOVERNMENT
BT1 administration

REGIONAL PLANNING
BT1 planning

REGIONS
NT1 administrative areas
NT1 agricultural regions
NT1 economic regions

REGRESSION ANALYSIS
BT1 statistical inference
BT2 statistical analysis
BT3 statistics

REGULATIONS
BT1 legislation

REGULATIONS (C)
BT2 legal sources
NT1 decrees
rt licenses and permits

REINVESTMENT
BT1 investment
BT2 agricultural situation

RELIGIONS
BT1 social systems
NT1 churches
NT2 churches in rural areas

REMOTE SENSING
NT1 satellite surveys

REMUNERATION
sf labour payment
sf payment schemes
BT1 labour economics
NT1 piece work wages
NT1 time wages
NT1 wage rates
NT1 wages
NT1 wages in kind

RENT
BT1 valuation
rt tenancy

REPAYMENT
sf debt servicing
BT1 credit policy
NT1 debt consolidation
NT1 interest repayments

REPLACEMENT VALUE
BT1 balance sheet analysis

REPLANTING
BT1 tillage
BT2 cultivation methods
rt planting

REPORTS
BT1 publications

RESEARCH
uf research and development
NT1 agricultural research
NT1 applied research
NT1 basic research
NT1 government research
NT1 private research
NT1 university research

research and development
USE **research**

RESEARCH ASSOCIATIONS
BT1 research institutions
BT2 organization of research

RESEARCH CONTRACTS
BT1 research policy
rt financing research

RESEARCH INSTITUTES
BT1 research institutions
BT2 organization of research

RESEARCH INSTITUTIONS
sf institutions
BT1 organization of research
NT1 experimental stations
NT1 laboratories
NT1 pilot farms
NT1 research associations
NT1 research institutes
rt rural sociology research

research organization
USE **organization of research**

RESEARCH PERSONNEL
BT1 organization of research
NT1 research teams

RESEARCH PERSONNEL (C)
NT1 research workers
NT2 agricultural economists
NT2 rural sociologists

RESEARCH POLICY
NT1 diffusion of research
NT1 evaluation of research
NT1 financing research
NT1 implementation of research
NT1 planning of research
NT2 research projects
NT1 research contracts
rt policy

RESEARCH PROJECTS
BT1 planning of research
BT2 research policy

RESEARCH TEAMS
BT1 research personnel
BT2 organization of research

RESEARCH WORKERS
BT1 research personnel
BT2 organization of research
NT1 agricultural economists
NT1 rural sociologists

RESERVE FUNDS
BT1 liabilities
BT2 balance sheets
BT3 balance sheet analysis
rt self finance

reserve stocks
USE **buffer stocks**

resettlement
USE **settlement**

RESOURCE ALLOCATION
BT1 agricultural situation

REST PERIODS
BT1 work study
BT2 labour economics

restrictives practices
USE **imperfect competition**

results
SEE **farm results**
OR **profitability**

RESUMPTION OF POSSESSION
BT1 termination of leases
rt agricultural tenure
rt landowners

RETAIL MARKETING
BT1 marketing channels
BT2 marketing
NT1 direct marketing
NT1 street markets
NT1 supermarkets

RETAIL PRICES
BT1 prices
rt consumer prices

RETIRED PEOPLE
BT1 non-gainfully employed
BT2 population structure
BT3 population
rt retirement

RETIREMENT
BT1 working conditions
rt retired people

RETRAINING
BT1 vocational training
BT2 education
rt continuing education
rt occupational change

RETROSPECTIVE SEARCHES
BT1 information retrieval

RETURNS
BT1 cost analysis

REUNION
BT1 southern africa
BT2 africa south of sahara
BT3 africa
rt france
rt french-speaking africa

REVALUTION
BT1 monetary parity
BT2 monetary policy

RHODESIA
uf *southern rhodesia*
BT1 southern africa
BT2 africa south of sahara
BT3 africa
rt english-speaking africa

RHUBARB
BT1 fruits

RICE
BT1 cereals
BT2 crops

RICE FLOUR
BT1 flour
BT2 cereal products

RIGHT OF ACCESS

rio de oro
USE **western sahara**

rio muni
USE **equatorial guinea**

RISK
BT1 investment planning
BT2 farm planning
rt uncertainty

river basins
USE **watersheds**

ROAD TRANSPORT
BT1 transport

ROADS
NT1 farm roads
NT1 rural roads

RODENTICIDES
BT1 pesticides

ROLE CHANGES
BT1 roles
BT2 social stratification
BT3 social structure

ROLE CONFLICTS
BT1 roles
BT2 social stratification
BT3 social structure

ROLES
BT1 social stratification
BT2 social structure
NT1 aspirations
NT1 role changes
NT1 role conflicts

ROLLING
BT1 tillage
BT2 cultivation methods

ROMANIA
uf *rumania*
BT1 eastern europe
BT2 europe
rt cmea

ROOT CROPS
BT1 crops
NT1 fodder beet
NT1 manioc
NT1 potatoes
NT2 feed potatoes
NT2 industrial potatoes
NT2 ware potatoes
NT1 sugar beet
NT1 turnips
NT1 yams

ROTATION
BT1 cultivation methods
rt catch crops

ROUGHAGE
BT1 farm produced fodder
BT2 feedingstuffs

RUBBER
BT1 industrial crops

RUBBER INDUSTRY
BT1 non-food industries

RULING CLASSES
BT1 social classes
BT2 social stratification
BT3 social structure

rumania
USE **romania**

RURAL ANIMATION
BT1 community development
BT2 rural development
BT3 rural environment
BT4 environment

RURAL AREAS
BT1 rural environment
BT2 environment
rt agricultural geography

RURAL COMMUNITIES
BT1 rural environment
BT2 environment
NT1 villages

RURAL DEPOPULATION
BT1 migration
rt occupational change

RURAL DEVELOPMENT
uf *integrated rural development*
BT1 rural environment
BT2 environment
NT1 community development
NT2 rural animation
NT1 settlement

RURAL ECONOMY
NT1 rural unemployment

RURAL ENVIRONMENT
BT1 environment
NT1 agricultural geography
NT1 rural areas
NT1 rural communities
NT2 villages
NT1 rural development
NT2 community development
NT3 rural animation
NT2 settlement

RURAL HOUSING
sf *buildings*
BT1 living conditions
NT1 farm dwellings

RURAL INDUSTRY
BT1 industry
rt rural planning

RURAL PLANNING
BT1 planning
NT1 building controls

RURAL PLANNING (C)
NT1 central places
rt infrastructure
rt rural industry

RURAL POPULATION
BT1 population structure
BT2 population
rt rural women

RURAL ROADS
BT1 roads

RURAL SOCIAL POLICY
BT1 social policy
rt policy
rt social security

RURAL SOCIETIES
BT1 societies

RURAL SOCIOLOGISTS
BT1 research workers
BT2 research personnel
BT3 organization of research

RURAL SOCIOLOGY
rt rural-urban migration

RURAL SOCIOLOGY RESEARCH
NT1 sociological analysis
NT1 sociological method
NT1 sociology
rt documentation
rt research institutions
rt social geography

RURAL UNEMPLOYMENT
BT1 rural economy

RURAL WELFARE
BT1 living conditions

RURAL WOMEN
BT1 women
rt rural population

RURAL YOUTH
BT1 youth
BT2 age
BT2 family size
BT3 families

RURAL-URBAN MIGRATION
BT1 agricultural development
BT1 geographical mobility
BT2 migration
rt rural sociology

RURAL-URBAN RELATIONS
BT1 environment
NT1 urbanization

russia
USE **ussr**

RWANDA
BT1 central africa
BT2 africa south of sahara
BT3 africa
BT1 east africa
rt french-speaking africa

RYE
BT1 cereals
BT2 crops

RYE FLOUR
BT1 flour
BT2 cereal products

sacks
USE **packs**

SAFER
(french group for cooperative farming: societe d'amenagement foncier et d'etablissement rural)
BT1 co-operative farm enterprises

SAFER (C)
rt land banks

SAFETY AT WORK
BT1 working conditions
NT1 accident prevention
NT1 industrial illness

salaries
USE **wages**

SALES
BT1 trade
rt marketing techniques
rt turnover

SALES PROMOTION
BT1 marketing techniques
NT1 labelling
NT2 quality labelling
NT3 standard labelling
NT2 trade marks
NT1 launching a product
NT1 product image
NT1 product presentation
NT1 publicity
NT2 advertizing
NT2 trade fairs
NT1 special offers

salvador
USE **el salvador**

SAMPLING
BT1 statistical analysis
BT2 statistics
rt data collection
rt surveys

SAN MARINO
BT1 mediterranean countries
BT1 western europe
BT2 europe

SAO TOME AND PRINCIPE
uf principe
BT1 central africa
BT2 africa south of sahara
BT3 africa
rt portugal

SATELLITE SURVEYS
BT1 remote sensing
rt aerial surveys

SAUDI ARABIA
BT1 arab countries
BT1 west asia
BT2 asia

SAUSAGES
BT1 meat products

savannah
USE **prairies**

SAVINGS
BT1 finance

SAVINGS BANKS
BT1 banks

SAVINGS FUNCTIONS
BT1 single equation analysis
BT2 parametric methods
BT3 econometrics

SAVOY CABBAGES
BT1 cabbages
BT2 vegetables

scale
USE **economies of scale**

SCANDINAVIA
uf northern europe
BT1 western europe
BT2 europe

SCANDINAVIA (C)
NT1 denmark
NT2 denmark (bornholm)
NT2 denmark (fyn)
NT2 denmark (jutland)
NT2 denmark (lolland-falster)
NT2 denmark (zealand)
NT1 faroe islands
NT1 finland
NT1 greenland
NT1 iceland
NT1 norway
NT1 sweden

scholarship
USE **educational grants**

SCHOOLS
BT1 education
NT1 primary education
NT1 secondary education

SDI SERVICES
BT1 information retrieval

SEA TRANSPORT
BT1 transport

SEASONAL FLUCTUATIONS
BT1 fluctuations
BT2 trade cycles
BT3 economic development

SEASONAL LABOUR
BT1 hired labour
BT2 agricultural manpower

SEASONAL MIGRATION
BT1 geographical mobility
BT2 migration

SEASONAL UNEMPLOYMENT
BT1 unemployment
BT2 employment

SEASONALITY
BT1 time series
BT2 dispersion and distribution
BT3 descriptive statistics
BT4 statistics

SECONDARY EDUCATION
BT1 schools
BT2 education

SECONDARY JOURNALS
BT1 periodicals
BT2 publications
NT1 bibliographies

SECONDARY SECTOR
sf sectors of production
BT1 economic sectors

SECTORAL DEVELOPMENT
BT1 economic development
rt agricultural development
rt economic sectors

SECTORAL PLANNING
BT1 planning

sectors of production
SEE **economic sectors**
OR **primary sector**
OR **secondary sector**
OR **tertiary sector**

SEED INDUSTRY
BT1 input industries
rt seeds

SEEDS
BT1 plant breeding
rt farm inputs
rt high yielding varieties
rt seed industry
rt sowing

SELECTIVE BREEDING
BT1 animal breeding
BT2 livestock sector

SELECTIVE HERBICIDES
BT1 pesticides
rt farm inputs

SELF ESTEEM
BT1 individuals

SELF FINANCE
uf *ploughback*
BT1 finance
rt reserve funds

SELF MANAGEMENT
BT1 management
BT2 human resources
BT3 economic resources
BT4 economic situation

SELF PERCEPTION
BT1 individuals

SELF SUFFICIENCY
BT1 supply balance
rt domestic production

SELLING PRICES
BT1 prices

SEMI-PRIVATE ORGANIZATIONS
uf *parastatal organizations*
BT1 administration

SEMINARS
BT1 teaching methods

SENEGAL
BT1 west africa
BT2 africa south of sahara
BT3 africa
rt french-speaking africa

sensory evaluation+
USE **consumer behaviour**
AND **quality**

SESAME
BT1 oilseeds
BT2 crops

SETTING-UP GRANTS
sf *grants*
BT1 subsidies
BT2 agricultural financial policy
rt adjustment of production

SETTLEMENT
uf *resettlement*
BT1 rural development
BT2 rural environment
BT3 environment

SEVERANCE ALLOWANCE
BT1 subsidies
BT2 agricultural financial policy
rt farm closures

SEXUAL BEHAVIOUR
BT1 families
rt birth rate
rt family planning

SEYCHELLES
BT1 east africa
BT2 africa south of sahara
BT3 africa
rt united kingdom

SHALLOTS
BT1 vegetables

share cropping
USE **metayage**

SHARE OF THE MARKET
BT1 supply balance
rt competition

SHEA NUTS
BT1 nuts

SHEEP
NT1 dual purpose sheep
NT1 ewes
NT1 lambs
NT1 rams
NT1 wool sheep breeds
rt livestock
rt sheep farming

SHEEP FARMING
BT1 livestock enterprises
BT2 farm enterprises
rt sheep

SHELLFISH
rt aquaculture

SHORT TERM CREDIT
BT1 credit
BT2 credit transactions

shrubs
USE **trees and shrubs**

SICKNESS BENEFITS
BT1 social insurance
BT2 social security
rt insurances

SIERRA LEONE
BT1 west africa
BT2 africa south of sahara
BT3 africa
rt commonwealth
rt english-speaking africa

SILAGE
BT1 farm produced fodder
BT2 feedingstuffs
rt preservation

SILK WORMS
BT1 livestock enterprises
BT2 farm enterprises

SILO MAIZE
BT1 maize
BT2 cereals
BT3 crops

SILOS
BT1 storage

SIMULATED FOODS
uf *imitation foods*
uf *synthetic foods*
NT1 artificial sweeteners
NT1 margarine
NT1 meat analogues
NT1 milk substitutes
rt substitution

SIMULATION
BT1 non-optimizing methods
BT2 operations research
NT1 monte carlo method

SIMULTANEOUS EQUATION ANALYSIS
BT1 parametric methods
BT2 econometrics

SINGAPORE
BT1 south east asia
BT2 asia
rt commonwealth

single cell proteins
USE **synthetic proteins**

SINGLE ENTRY BOOK-KEEPING
BT1 book-keeping systems

SINGLE ENTRY BOOK-KEEPING (C)
BT2 book-keeping

SINGLE EQUATION ANALYSIS
BT1 parametric methods
BT2 econometrics
NT1 consumption functions
NT1 demand functions
NT1 elasticities
NT1 flexibility
NT1 investment functions
NT1 production functions
NT1 savings functions
NT1 supply functions
NT1 utility functions

SISAL
BT1 fibre crops
BT2 crops

SKILLED LABOUR
BT1 hired labour
BT2 agricultural manpower

SKIM MILK
BT1 milk

slaughter animals
USE **livestock**

SLAUGHTER PREMIUMS
BT1 subsidies
BT2 agricultural financial policy

slaughterhouses
USE **abattoirs**

slimming foods
USE **prepared foods**

sludge
USE **slurry**

SLUICEGATE PRICES
BT1 prices

SLURRY
uf *sludge*
BT1 organic fertilizers
BT2 fertilizers

SLURRY REMOVAL
uf *mucking out*
BT1 livestock husbandry
BT2 livestock sector
rt pollution by agriculture
rt waste disposal

SMALL ANIMAL REARING
BT1 livestock enterprises
BT2 farm enterprises
rt rabbits

SMALL FARMS
BT1 types of farming
BT2 agricultural systems
rt farm size

SMALL GROUPS
BT1 group size
BT2 groups
NT1 neighbourhood
rt families

smoked bacon
USE **bacon**

SNAILS

SNOW
BT1 precipitation
BT2 climatic factors

SOCIAL ACTIVITY
BT1 social participation
BT2 socialization

SOCIAL ADJUSTMENT
BT1 socialization
NT1 maladjustment

SOCIAL ANTHROPOLOGY
NT1 ethnic groups
rt culture

SOCIAL CHANGE
NT1 modernization
 NT2 adoption of innovations
 NT2 industrialization
 NT2 technical progress
  NT3 appropriate technology
  NT3 technology transfers
NT1 social development
rt change
rt economic development

SOCIAL CLASSES
BT1 social stratification
 BT2 social structure
NT1 middle classes
NT1 ruling classes
NT1 upper classes
NT1 working classes

social conditions
USE **living conditions**

SOCIAL CONSCIOUSNESS
BT1 socialization

SOCIAL CRISIS
BT1 socialization
NT1 conflict
NT1 deviant behaviour
 NT2 alcoholism
 NT2 criminality
NT1 social isolation
NT1 social unrest
 NT2 class conflict
rt economic depression
rt maladjustment
rt work conflicts

SOCIAL CUSTOMS
BT1 social standards
 BT2 social systems

SOCIAL DEVELOPMENT
BT1 social change

SOCIAL DIFFERENTIATION
BT1 socialization

SOCIAL EQUILIBRIUM
BT1 social structure

SOCIAL GEOGRAPHY
rt agricultural geography
rt geography
rt rural sociology research

SOCIAL INSTITUTIONS
sf institutions
BT1 social systems
NT1 institution building

SOCIAL INSURANCE
uf national insurance
BT1 social security
NT1 invalid benefits
NT1 sickness benefits
rt insurances

SOCIAL INTEGRATION
BT1 socialization

SOCIAL ISOLATION
BT1 social crisis
 BT2 socialization

SOCIAL LEGISLATION
NT1 labour legislation
 NT2 labour protection
 NT2 strikes
rt contract legislation

SOCIAL MOBILITY
sf mobility
BT1 social stratification
 BT2 social structure
NT1 horizontal mobility
NT1 vertical mobility
 NT2 downward mobility
 NT2 upward mobility

social order
USE **social structure**

SOCIAL PARTICIPATION
BT1 socialization
NT1 social activity

SOCIAL POLICY
NT1 rural social policy
rt legislation
rt policy
rt social security

SOCIAL SANCTIONS
BT1 social systems

SOCIAL SECURITY
NT1 accident benefits
NT1 family allowances
NT1 health services
NT1 life insurance
NT1 maternity benefits
NT1 old age benefits
NT1 social insurance
 NT2 invalid benefits
 NT2 sickness benefits
NT1 social services
NT1 unemployment insurance
rt insurances
rt legal systems
rt rural social policy
rt social policy

SOCIAL SERVICES
BT1 social security
rt insurances

SOCIAL STANDARDS
BT1 social systems
NT1 social customs
NT1 traditions
rt standards

SOCIAL STRATIFICATION
BT1 social structure
NT1 ownership conditions
NT1 roles
 NT2 aspirations
 NT2 role changes
 NT2 role conflicts
NT1 social classes
 NT2 middle classes
 NT2 ruling classes
 NT2 upper classes
 NT2 working classes
NT1 social mobility
 NT2 horizontal mobility
 NT2 vertical mobility
  NT3 downward mobility
  NT3 upward mobility
NT1 status
 NT2 elites
 NT2 occupational prestige
 NT2 occupational status
 NT2 socioeconomic status

SOCIAL STRUCTURE
uf social order
NT1 social equilibrium
NT1 social stratification
 NT2 ownership conditions
 NT2 roles
  NT3 aspirations
  NT3 role changes
  NT3 role conflicts
 NT2 social classes
  NT3 middle classes
  NT3 ruling classes
  NT3 upper classes

SOCIAL STRUCTURE (C)
  NT3 working classes
 NT2 social mobility
  NT3 horizontal mobility
  NT3 vertical mobility
   NT4 downward mobility
   NT4 upward mobility
 NT2 status
  NT3 elites
  NT3 occupational prestige
  NT3 occupational status
  NT3 socioeconomic status
rt structure

SOCIAL SYSTEMS
NT1 legal systems
 NT2 legal rights
NT1 power
 NT2 authority
 NT2 leadership
 NT2 political power
NT1 religions
 NT2 churches
  NT3 churches in rural areas
NT1 social institutions
 NT2 institution building
NT1 social sanctions
NT1 social standards
 NT2 social customs
 NT2 traditions

SOCIAL UNREST
BT1 social crisis
 BT2 socialization
NT1 class conflict

SOCIALISM
BT1 political systems

SOCIALIZATION
(sociology)
NT1 collectivization
NT1 social adjustment
 NT2 maladjustment
NT1 social consciousness
NT1 social crisis
 NT2 conflict
 NT2 deviant behaviour
  NT3 alcoholism
  NT3 criminality
 NT2 social isolation
 NT2 social unrest
  NT3 class conflict
NT1 social differentiation
NT1 social integration
NT1 social participation
 NT2 social activity
rt co-operation

SOCIETIES
NT1 agricultural society
 NT2 peasantry
NT1 industrial societies
NT1 mass society
NT1 pastoral society
 NT2 nomads
NT1 plural societies
NT1 rural societies
NT1 urban society

SOCIOECONOMIC ORGANIZATION
NT1 centralisation
NT1 co-operation
NT1 concentration
 NT2 farm amalgamation
NT1 coordination
NT1 decentralization
NT1 diversification
NT1 integration
 NT2 horizontal integration
 NT2 vertical integration
NT1 rationalization
NT1 specialization

SOCIOECONOMIC STATUS
BT1 status
 BT2 social stratification
  BT3 social structure

SOCIOECONOMIC STATUS (C)
rt living conditions

SOCIOLOGICAL ANALYSIS
BT1 rural sociology research

SOCIOLOGICAL METHOD
BT1 rural sociology research
rt econometrics
rt operations research

SOCIOLOGY
BT1 rural sociology research

SOCIOLOGY OF WORK
rt work study
rt working conditions

SOFT CHEESE
BT1 cheese
BT2 milk products

SOFT DRINKS
BT1 beverages
NT1 coffee
NT1 mate
NT1 tea
rt fruit juices

SOFT WHEAT
BT1 wheat
BT2 cereals
BT3 crops

SOIL
rt soil analysis
rt soil properties

SOIL ANALYSIS
rt soil
rt soil properties

SOIL CONSERVATION
uf *erosion*
BT1 land improvement
BT2 cultivation methods

SOIL FERTILITY
uf *fertility*
BT1 soil properties

SOIL HUMIDITY
BT1 soil properties

SOIL PROPERTIES
NT1 soil fertility
NT1 soil humidity
rt soil
rt soil analysis

soilless culture
USE **hydroponics**

SOLAR ENERGY
BT1 energy
BT2 economic resources
BT3 economic situation

soluble coffee
USE **instant coffee**

SOMALIA
BT1 east africa
BT2 africa south of sahara
BT3 africa

SONS
BT1 children
BT2 family size
BT3 families

SORGHUM
uf *kaffir corn*
BT1 cereals
BT2 crops

SOUTH AFRICA
uf *south african republic*

SOUTH AFRICA (C)
BT1 southern africa
BT2 africa south of sahara
BT3 africa
rt english-speaking africa

south african republic
USE **south africa**

SOUTH AMERICA
sf *latin america*
BT1 america
NT1 argentina
NT1 bolivia
NT1 brazil
NT1 chile
NT1 colombia
NT1 ecuador
NT1 falkland islands
NT1 french guiana
NT1 guyana
NT1 paraguay
NT1 peru
NT1 surinam
NT1 uruguay
NT1 venezuela
rt caribbean
rt central america

SOUTH EAST ASIA
BT1 asia
NT1 bangladesh
NT1 bhutan
NT1 brunei
NT1 burma
NT1 india
NT1 indochina
NT2 kampuchea
NT2 lao
NT2 vietnam
NT1 indonesia
NT1 malaysia
NT1 maldives
NT1 nepal
NT1 pakistan
NT1 philippines
NT1 portuguese timor
NT1 singapore
NT1 sri lanka
NT1 thailand

south korea
USE **korea r**

south sandwich islands
USE **falkland islands**

south vietnam
USE **vietnam**

south west africa
USE **namibia**

SOUTHERN AFRICA
BT1 africa south of sahara
BT2 africa
NT1 angola
NT1 botswana
NT1 comoro islands
NT1 lesotho
NT1 madagascar
NT1 malawi
NT1 mauritius
NT1 mozambique
NT1 namibia
NT1 reunion
NT1 rhodesia
NT1 south africa
NT1 st helena
NT1 swaziland
NT1 zambia

southern europe
USE **mediterranean countries**

southern rhodesia
USE **rhodesia**

soviet union
USE **ussr**

SOWING
BT1 tillage
BT2 cultivation methods
rt seeds

SOWS
BT1 pigs

SOY OILS
uf *soybean oils*
BT1 vegetable oils
BT2 oils and fats

SOY PRODUCTS
BT1 vegetable products

soybean oils
USE **soy oils**

SOYBEANS
BT1 oilseeds
BT2 crops

SPAIN
BT1 mediterranean countries
BT1 western europe
BT2 europe
rt andorra
rt canary islands

spanish guinea
USE **equatorial guinea**

spanish sahara
USE **western sahara**

SPECIAL COSTS
BT1 production costs
BT2 costs
BT3 cost analysis

SPECIAL OFFERS
BT1 sales promotion
BT2 marketing techniques

SPECIALIZATION
BT1 socioeconomic organization
rt specialized farming

SPECIALIZED FARMING
BT1 types of farming
BT2 agricultural systems
rt specialization

SPECIALIZED HIGHER EDUCATION
sf *higher education*
BT1 education

SPECTRAL ANALYSIS
BT1 statistical inference
BT2 statistical analysis
BT3 statistics

SPICE PLANTS
BT1 crops

SPINACH
BT1 vegetables

SPIRITS
(excluding liqueurs)
BT1 alcoholic beverages
BT2 beverages

SPORT
BT1 leisure

SPRAY IRRIGATION
BT1 land improvement
BT2 cultivation methods

SPRAYING
BT1 plant protection

SQUASHES
BT1 vegetables

SRI LANKA
uf *ceylon*
BT1 south east asia
BT2 asia
rt commonwealth

ST HELENA
(including dependencies)
BT1 southern africa
BT2 africa south of sahara
BT3 africa
BT1 west africa
rt english-speaking africa
rt united kingdom

ST KITTS AND NEVIS AND ANGUILLA
uf *anguilla, nevis*
BT1 west indies associated states
BT2 caribbean
BT3 america

ST LUCIA
BT1 west indies associated states
BT2 caribbean
BT3 america

ST PIERRE AND MIQUELON
BT1 north america
BT2 america
rt france

ST VINCENT
BT1 west indies associated states
BT2 caribbean
BT3 america

STABILIZATION
BT1 trade cycles
BT2 economic development

STABLING
BT1 livestock husbandry
BT2 livestock sector

STAGNATION
BT1 trade cycles
BT2 economic development

STALLIONS
BT1 horses

STAMP DUTY
uf *property transfer tax*
BT1 indirect taxation
BT2 fiscal policy
BT3 public finance

STANDARD DEVIATION
BT1 dispersion and distribution
BT2 descriptive statistics
BT3 statistics

STANDARD LABELLING
BT1 quality labelling
BT2 labelling
BT3 sales promotion
BT4 marketing techniques

standard of living
USE **living standards**

STANDARDS
NT1 date stamping
NT1 grading
NT1 quality standards
NT1 quantity controls
rt social standards

STARCH INDUSTRY
BT1 food industry
rt pasta industry

starch sugar
USE **isomerose**

STATE FARMS
BT1 farms

state ownership
USE **public ownership**

STATISTICAL ANALYSIS
sf *analysis*
BT1 statistics
NT1 sampling
NT1 statistical inference
NT2 chi square test
NT2 component analysis
NT2 correlation analysis
NT2 cross sectional analysis
NT2 discriminant analysis
NT2 factorial analysis
NT2 least squares method
NT2 multicollinearity
NT2 regression analysis
NT2 spectral analysis
NT2 student test
NT2 variance analysis
rt stochastic processes

STATISTICAL INFERENCE
BT1 statistical analysis
BT2 statistics
NT1 chi square test
NT1 component analysis
NT1 correlation analysis
NT1 cross sectional analysis
NT1 discriminant analysis
NT1 factorial analysis
NT1 least squares method
NT1 multicollinearity
NT1 regression analysis
NT1 spectral analysis
NT1 student test
NT1 variance analysis
rt game theory

STATISTICS
NT1 descriptive statistics
NT2 central value characteristics
NT3 averages
NT2 dispersion and distribution
NT3 standard deviation
NT3 time series
NT4 seasonality
NT3 variance
NT1 graphs
NT1 statistical analysis
NT2 sampling
NT2 statistical inference
NT3 chi square test
NT3 component analysis
NT3 correlation analysis
NT3 cross sectional analysis
NT3 discriminant analysis
NT3 factorial analysis
NT3 least squares method
NT3 multicollinearity
NT3 regression analysis
NT3 spectral analysis
NT3 student test
NT3 variance analysis
NT1 stochastic processes
NT2 probability analysis
NT1 theory of graphs
rt censuses

STATUS
BT1 social stratification
BT2 social structure
NT1 elites
NT1 occupational prestige
NT1 occupational status
NT1 socioeconomic status

statutes
USE **laws**

steers
USE **bullocks**

steppe
USE **prairies**

STEREOTYPES
BT1 behaviour

STOCHASTIC PROCESSES
BT1 statistics
NT1 probability analysis
rt statistical analysis

STOCHASTIC PROGRAMMING
BT1 programming

STOCK ACCOUNTING
BT1 book-keeping

STOCK-PILING SUBSIDIES
BT1 subsidies
BT2 agricultural financial policy
rt stocks

STOCKS
NT1 buffer stocks
rt stock-piling subsidies
rt storage
rt supply balance
rt surpluses

STONE FRUITS
BT1 fruits
NT1 apricots
NT1 cherries
NT1 olives
NT1 peaches
NT2 nectarines
NT1 plums

STORAGE
NT1 bulk storage
NT1 cold storage
NT1 controlled atmosphere storage
NT1 silos
NT1 storage tanks
NT1 warehouses
rt handling
rt stocks

STORAGE TANKS
BT1 storage

storm insurance
USE **crop insurance**

STRAW
BT1 farm produced fodder
BT2 feedingstuffs
rt crop residues

STRAWBERRIES
BT1 berries
BT2 fruits

STREET MARKETS
BT1 retail marketing
BT2 marketing channels
BT3 marketing

STRIKES
BT1 labour legislation
BT2 social legislation

STRUCTURAL POLICY
uf *common guidance policy*
BT1 agricultural policy
NT1 land consolidation
rt acreage subsidies
rt agrarian reform
rt farm closures
rt farm plans
rt less favoured areas
rt physical planning
rt policy
rt production policy

STRUCTURAL UNEMPLOYMENT
BT1 unemployment
BT2 employment

STRUCTURE
(use only for types of structure not separatly

STRUCTURE (C)
listed)
rt agricultural structure
rt family structure
rt farm structure
rt population structure
rt production structure
rt social structure

STUDENT TEST
BT1 statistical inference
BT2 statistical analysis
BT3 statistics

STUDENTS
uf *pupils*
rt teachers

STUDY LEAVE
BT1 educational policy

SUBNATIONAL LEGISLATION
BT1 laws
BT2 legislation
BT3 legal sources

SUBSIDIES
sf *grants*
BT1 agricultural financial policy
NT1 acreage subsidies
NT1 grubbing up grants
NT1 improvement grants
NT1 non-marketing premiums
NT1 setting-up grants
NT1 severance allowance
NT1 slaughter premiums
NT1 stock-piling subsidies
rt agricultural disasters
rt damage
rt support measures

SUBSISTENCE FARMING
uf *traditional farming*
BT1 agricultural systems

SUBSOILING
BT1 tillage
BT2 cultivation methods

SUBSTITUTION
rt elasticities
rt simulated foods

SUBURBAN AGRICULTURE
BT1 urban areas
BT2 urban environment
BT3 environment

SUCCESSION
BT1 acquisition of ownership
NT1 indemnification
NT1 indivisible estates
NT1 partition of estates
rt inheritance

SUDAN
BT1 arab countries
BT1 east africa
BT2 africa south of sahara
BT3 africa
rt english-speaking africa

SUGAR
NT1 beet sugar
NT1 cane sugar
NT1 isomerose
rt artificial sweeteners
rt food
rt sugar beet
rt sugar cane
rt sugar confectionery

SUGAR BEET
BT1 root crops
BT2 crops
rt beet sugar
rt sugar

SUGAR CANE
BT1 crops
rt cane sugar
rt sugar

SUGAR CONFECTIONERY
rt chocolate
rt food
rt sugar

SUGAR INDUSTRY
BT1 food industry

SUNFLOWER SEED OILS
BT1 vegetable oils
BT2 oils and fats

SUPERMARKETS
BT1 retail marketing
BT2 marketing channels
BT3 marketing

supervised credit+
USE **agricultural credit**
AND **extension**

SUPPLY
BT1 market economics
rt production
rt quotas
rt supply balance

SUPPLY AND DEMAND
BT1 market economics
rt demand
rt supply balance

SUPPLY BALANCE
NT1 food supply
NT1 self sufficiency
NT1 share of the market
rt balance of trade
rt demand
rt stocks
rt supply
rt supply and demand

supply co-operatives
USE **co-operative purchasing**

SUPPLY FUNCTIONS
BT1 single equation analysis
BT2 parametric methods
BT3 econometrics

SUPPORT MEASURES
sf *grants*
BT1 agricultural policy
rt acreage subsidies
rt policy
rt subsidies

SURINAM
BT1 south america
BT2 america
rt netherlands

SURPLUSES
uf *overproduction*
BT1 production controls
BT2 production policy
rt quotas
rt stocks

SURVEYS
NT1 farm surveys
NT2 european farm accounting network
rt censuses
rt data collection
rt sampling

SWAZILAND
BT1 southern africa
BT2 africa south of sahara
BT3 africa
rt commonwealth
rt english-speaking africa

SWEDEN
BT1 scandinavia
BT2 western europe
BT3 europe
rt efta

SWEDES
BT1 vegetables

SWEET ALMONDS
BT1 almonds
BT2 nuts

SWEET CORN
BT1 vegetables

SWEET POTATOES
BT1 vegetables

SWEETENED CONDENSED MILK
sf *condensed milk*
BT1 milk products

sweeteners
USE **artificial sweeteners**

SWITZERLAND
BT1 western europe
BT2 europe
rt efta

synthetic foods
USE **simulated foods**

SYNTHETIC PROTEINS
uf *single cell proteins*
BT1 proteins

SYRIA
uf *syrian arab republic*
BT1 arab countries
BT1 mediterranean countries
BT1 west asia
BT2 asia

syrian arab republic
USE **syria**

SYSTEMS ANALYSIS
BT1 non-optimizing methods
BT2 operations research
rt mathematics

TABLE WINE
BT1 wine
BT2 alcoholic beverages
BT3 beverages

taiwan
USE **china**

TANZANIA
uf *united republic of tanzania*
BT1 east africa
BT2 africa south of sahara
BT3 africa
rt english-speaking africa

tapioca
USE **manioc**

TARGET PRICES
BT1 prices
rt agricultural prices
rt price policy

TARIFFS
uf *customs duties*
NT1 preferential tariffs
rt indirect taxation
rt trade barriers
rt variable levies

TAX AGREEMENTS
BT1 non-tariff barriers to trade

TAXABLE RENT VALUE
BT1 valuation

taxation
SEE **capital taxation**
OR **direct taxation**
OR **indirect taxation**

taxation policy
USE **fiscal policy**

TAXES
BT1 fiscal policy
BT2 public finance

tchad
USE **chad**

TEA
uf *black tea*
uf *decaffeinated tea*
uf *fermented tea*
uf *tea blends*
uf *tea leaf*
BT1 soft drinks
BT2 beverages
rt instant tea

tea blends
USE **tea**

TEA INDUSTRY
BT1 beverage industry
BT2 food industry

tea leaf
USE **tea**

tea powder
USE **instant tea**

TEACHERS
BT1 occupations
rt education
rt students

TEACHING METHODS
NT1 audiovisual aids
NT2 educational radio
NT2 educational television
NT1 case studies
NT1 discussion groups
NT1 management games
NT1 programmed learning
NT1 seminars

TECHNICAL AID
sf *aid*
BT1 development aid
rt technology transfers

TECHNICAL PROGRESS
BT1 modernization
BT2 social change
NT1 appropriate technology
NT1 technology transfers
rt agricultural engineering
rt automation

TECHNOLOGY TRANSFERS
BT1 technical progress
BT2 modernization
BT3 social change
rt technical aid

teleprocessing
USE **data processing**

TELEVISION
BT1 mass media

TEMPERATE ZONES
BT1 climatic areas

TEMPEST
BT1 agricultural disasters

TEMPORARY GRASSLANDS
BT1 pastures
BT2 land

TENANCY
sf *leasing*
BT1 tenure systems
NT1 leases
rt rent

tenancy agreements
USE **farm leases**

tenant farmers
USE **tenants**

TENANTS
uf *tenant farmers*
rt agricultural tenure
rt landowners
rt tenants rights

TENANTS RIGHTS
rt tenants

TENURE SYSTEMS
uf *land tenure*
NT1 metayage
NT1 owner occupancy
NT1 tenancy
NT2 leases

TERMINATION OF LEASES
NT1 resumption of possession
rt option to purchase

TERMS OF TRADE
uf *comparative advantage*
BT1 balance of payments

TERRACING
BT1 cropping systems

TERTIARY SECTOR
sf *sectors of production*
BT1 economic sectors
NT1 professional services
NT2 veterinary services
NT1 public services

TEXTILE INDUSTRIES
BT1 non-food industries
NT1 cotton industry
NT2 cotton ginning
NT1 wool industry

textured vegetable proteins(tvp)
USE **meat analogues**

THAILAND
BT1 south east asia
BT2 asia

theory
SEE **bayesian theory**
OR **development theory**
OR **economic theories**
OR **equilibrium theory**
OR **growth theory**
OR **location theory**
OR **queueing theory**

THEORY OF GRAPHS
BT1 statistics
rt mathematics
rt modern mathematics
rt operations research

THESAURI
BT1 documentation languages
BT2 information processing

THESIS
BT1 publications

THINKING
BT1 decision making

THINNING
BT1 plant husbandry
BT2 cultivation methods

third world
USE **developing countries**

THRESHOLD PRICES
BT1 prices

TILLAGE
BT1 cultivation methods
NT1 harrowing
NT1 minimal tillage systems
NT1 no-tillage system
NT1 planting
NT1 ploughing
NT1 replanting
NT1 rolling
NT1 sowing
NT1 subsoiling

timbers
USE **forest products**

TIME SERIES
BT1 dispersion and distribution
BT2 descriptive statistics
BT3 statistics
NT1 seasonality

time study
USE **work study**

TIME WAGES
uf *hourly wages*
BT1 remuneration
BT2 labour economics

TOBACCO
BT1 industrial crops

TOBACCO INDUSTRY
BT1 non-food industries

tobago
USE **trinidad and tobago**

TOGO
BT1 west africa
BT2 africa south of sahara
BT3 africa
rt french-speaking africa

TOMATOES
BT1 vegetables

TONGA
BT1 oceania
rt commonwealth
rt united kingdom

TOOLS
BT1 farm equipment

TOPOGRAPHY
BT1 geography
rt aerial surveys

TORTOLA
uf *british virgin islands*
BT1 caribbean
BT2 america
rt united kingdom

TOTAL AGRICULTURAL EXPENDITURE
BT1 aggregate data

TOTAL COSTS
BT1 production costs
BT2 costs
BT3 cost analysis

TOURISM
uf *farm recreations*
NT1 farm holidays
rt ancillary enterprises

TOWNS
BT1 urban areas
BT2 urban environment
BT3 environment

TRACTORS
BT1 farm machinery
BT2 farm equipment

TRADE
NT1 barter
NT1 domestic trade
NT1 external trade
NT2 agricultural trade
NT2 exports
NT2 imports
NT1 international trade
NT1 intra-community trade
NT1 purchasing
NT1 sales
NT1 transit trade
NT1 turnover
rt agricultural trade
rt commercial law

TRADE AGREEMENTS
BT1 trade policy
NT1 international agreements
NT1 trade preferences

TRADE ASSOCIATIONS
sf associations
BT1 interest groups
BT2 groups
rt congresses

TRADE BARRIERS
BT1 protection
NT1 export controls
rt import controls
rt non-tariff barriers to trade
rt tariffs

TRADE CYCLES
BT1 economic development
NT1 economic depression
NT1 economic recovery
NT1 fluctuations
NT2 cyclic fluctuations
NT2 seasonal fluctuations
NT2 trends
NT1 stabilization
NT1 stagnation

TRADE FAIRS
BT1 publicity
BT2 sales promotion
BT3 marketing techniques

TRADE LIBERALIZATION
uf free trade policy
BT1 trade policy

TRADE MARKS
sf brands
BT1 labelling
BT2 sales promotion
BT3 marketing techniques
rt quality labelling

TRADE NEGOTIATIONS
BT1 trade policy
NT1 gatt
NT1 unctad

TRADE POLICY
NT1 balance of trade
NT1 commercial credit
NT2 capital leasing
NT1 export diversification
NT1 export promotion
NT2 export credits
NT2 export refunds
NT2 export subsidies
NT1 imports promotion
NT1 marketing boards
NT1 trade agreements
NT2 international agreements
NT2 trade preferences
NT1 trade liberalization
NT1 trade negotiations
NT2 gatt
NT2 unctad

TRADE POLICY (C)
NT1 trade relations
rt policy

TRADE PREFERENCES
BT1 trade agreements
BT2 trade policy

TRADE RELATIONS
BT1 trade policy

TRADE UNIONS
BT1 interest groups
BT2 groups
rt agricultural trade unions
rt labour protection

traditional farming
USE **subsistence farming**

TRADITIONS
BT1 social standards
BT2 social systems

training
USE **vocational training**

TRAINING LEVIES
BT1 educational policy

TRANSHUMANCE
BT1 livestock husbandry
BT2 livestock sector
rt nomads

TRANSIT TRADE
BT1 trade

TRANSITIONAL FARMING
(transition between subsistence and commercial farming)
BT1 agricultural systems

TRANSPORT
NT1 air transport
NT1 collection
NT1 container transport
NT1 inland transport
NT1 inland waterways
NT1 international transport
NT1 rail transport
NT1 road transport
NT1 sea transport
rt handling

TRANSPORT MODELS
BT1 models
BT1 optimization methods
BT2 operations research

TREATY OF ROME
BT1 ec laws
BT2 legal sources
NT1 ec regulations
rt ec legal decisions

TREE NURSERIES
BT1 horticulture
BT2 crop enterprises
BT3 farm enterprises
rt trees and shrubs

TREES AND SHRUBS
uf shrubs
BT1 crops
rt tree nurseries

TRENDS
BT1 fluctuations
BT2 trade cycles
BT3 economic development
rt change

TRINIDAD AND TOBAGO
uf tobago
BT1 caribbean
BT2 america
rt commonwealth

TRITICALE
BT1 cereals
BT2 crops
rt high yielding varieties

TROPICAL FRUITS
BT1 fruits
NT1 avocadoes
NT1 bananas
NT1 dates
NT1 guavas
NT1 mangoes
NT1 passion fruit
NT1 pawpaws
NT1 pineapples

TROPICS
BT1 climatic areas

trucial states
USE **united arab emirates**

TUNISIA
sf africa (maghreb)
BT1 arab countries
BT1 mediterranean countries
BT1 north africa
BT2 africa
rt french-speaking africa

TURKEY
BT1 mediterranean countries
BT1 western europe
BT2 europe

TURKEYS
BT1 poultry

TURKS AND CAICOS ISLANDS
uf caicos
BT1 caribbean
BT2 america
rt united kingdom

TURNIPS
BT1 root crops
BT2 crops

TURNOVER
BT1 trade
rt purchasing
rt sales

TWO GENERATION FAMILIES
BT1 nuclear families
BT2 family structure

TYPES OF FARMING
BT1 agricultural systems
NT1 collective farming
NT1 domestic gardens
NT1 family farms
NT1 full-time farming
NT1 large farms
NT1 medium sized farms
NT1 mixed farming
NT1 part-time farming
NT2 peasant workers
NT1 plantations
NT1 small farms
NT1 specialized farming
rt farms

UGANDA
BT1 east africa
BT2 africa south of sahara
BT3 africa
rt commonwealth
rt english-speaking africa

ujamaa villages
USE **communes**

uk
USE **united kingdom**

UKRAINIAN SSR
BT1 eastern europe

UKRAINIAN SSR (C)
BT2 europe
rt ussr

UN
(united nations)
BT1 organizations

UNCERTAINTY
BT1 investment planning
BT2 farm planning
rt risk

UNCTAD
(united nations conference on trade and development)
BT1 trade negotiations
BT2 trade policy

underdeveloped countries
USE **developing countries**

UNDEREMPLOYMENT
BT1 employment

UNEMPLOYED
BT1 non-gainfully employed
BT2 population structure
BT3 population

UNEMPLOYMENT
BT1 employment
NT1 seasonal unemployment
NT1 structural unemployment

UNEMPLOYMENT INSURANCE
BT1 social security
rt insurances

UNESCO
(united nations educational, scientific and cultural organization)
BT1 organizations

union of soviet socialist republics
USE **ussr**

UNITED ARAB EMIRATES
uf trucial states
BT1 persian gulf states
BT2 asia

united arab republic
USE **egypt**

UNITED KINGDOM
uf uk
BT1 western europe
BT2 europe
NT1 united kingdom (channel islands)
NT1 united kingdom (england and wales)
NT2 united kingdom (e.england)
NT2 united kingdom (e.midlands)
NT2 united kingdom (n.england)
NT2 united kingdom (s.w.england)
NT2 united kingdom (w.midlands)
NT2 united kingdom (wales)
NT2 united kingdom (yorks/lancs)
NT1 united kingdom (isle of man)
NT1 united kingdom (n.ireland)
NT1 united kingdom (scotland)
NT2 united kingdom (e.scotland)
NT2 united kingdom (highlands+islands)
NT2 united kingdom (n.scotland)
NT2 united kingdom (w.scotland)
rt antarctica
rt bahamas
rt belize
rt bermuda
rt brunei
rt cayman islands
rt commonwealth
rt european communities
rt falkland islands
rt gibraltar
rt hong kong
rt montserrat
rt pacific islands gb

UNITED KINGDOM (C)
rt seychelles
rt st helena
rt tonga
rt tortola
rt turks and caicos islands
rt west indies associated states

UNITED KINGDOM (CHANNEL ISLANDS)
BT1 united kingdom
BT2 western europe
BT3 europe

UNITED KINGDOM (E.ENGLAND)
BT1 united kingdom (england and wales)
BT2 united kingdom
BT3 western europe
BT4 europe

UNITED KINGDOM (E.MIDLANDS)
BT1 united kingdom (england and wales)
BT2 united kingdom
BT3 western europe
BT4 europe

UNITED KINGDOM (E.SCOTLAND)
BT1 united kingdom (scotland)
BT2 united kingdom
BT3 western europe
BT4 europe

UNITED KINGDOM (ENGLAND AND WALES)
BT1 united kingdom
BT2 western europe
BT3 europe
NT1 united kingdom (e.england)
NT1 united kingdom (e.midlands)
NT1 united kingdom (n.england)
NT1 united kingdom (s.w.england)
NT1 united kingdom (w.midlands)
NT1 united kingdom (wales)
NT1 united kingdom (yorks/lancs)

UNITED KINGDOM (HIGHLANDS+ISLANDS)
BT1 united kingdom (scotland)
BT2 united kingdom
BT3 western europe
BT4 europe

UNITED KINGDOM (ISLE OF MAN)
BT1 united kingdom
BT2 western europe
BT3 europe

UNITED KINGDOM (N.ENGLAND)
BT1 united kingdom (england and wales)
BT2 united kingdom
BT3 western europe
BT4 europe

UNITED KINGDOM (N.IRELAND)
uf united kingdom (ulster)
BT1 united kingdom
BT2 western europe
BT3 europe

UNITED KINGDOM (N.SCOTLAND)
BT1 united kingdom (scotland)
BT2 united kingdom
BT3 western europe
BT4 europe

UNITED KINGDOM (S.W.ENGLAND)
BT1 united kingdom (england and wales)
BT2 united kingdom
BT3 western europe
BT4 europe

UNITED KINGDOM (SCOTLAND)
BT1 united kingdom
BT2 western europe
BT3 europe
NT1 united kingdom (e.scotland)
NT1 united kingdom (highlands+islands)
NT1 united kingdom (n.scotland)
NT1 united kingdom (w.scotland)

united kingdom (ulster)
USE **united kingdom (n.ireland)**

UNITED KINGDOM (W.MIDLANDS)
BT1 united kingdom (england and wales)
BT2 united kingdom
BT3 western europe
BT4 europe

UNITED KINGDOM (W.SCOTLAND)
BT1 united kingdom (scotland)
BT2 united kingdom
BT3 western europe
BT4 europe

UNITED KINGDOM (WALES)
BT1 united kingdom (england and wales)
BT2 united kingdom
BT3 western europe
BT4 europe

UNITED KINGDOM (YORKS/LANCS)
BT1 united kingdom (england and wales)
BT2 united kingdom
BT3 western europe
BT4 europe

united mexican states
USE **mexico**

united republic of tanzania
USE **tanzania**

united states of america
USE **usa**

UNITED STATES VIRGIN ISLANDS
uf virgin islands us
BT1 caribbean
BT2 america

UNIVERSITIES
sf higher education
BT1 education

UNIVERSITY RESEARCH
BT1 research

UNOCCUPIED TIME
BT1 work study
BT2 labour economics

UNSKILLED LABOUR
BT1 hired labour
BT2 agricultural manpower

UPPER CLASSES
BT1 social classes
BT2 social stratification
BT3 social structure

UPPER VOLTA
BT1 west africa
BT2 africa south of sahara
BT3 africa
rt french-speaking africa

UPWARD MOBILITY
BT1 vertical mobility
BT2 social mobility
BT3 social stratification
BT4 social structure

URBAN AREAS
BT1 urban environment
BT2 environment
NT1 suburban agriculture
NT1 towns
NT1 urban hinterland

URBAN ENVIRONMENT
BT1 environment
NT1 urban areas
NT2 suburban agriculture
NT2 towns
NT2 urban hinterland

URBAN HINTERLAND
BT1 urban areas
BT2 urban environment
BT3 environment

URBAN POPULATION
BT1 population structure
BT2 population

URBAN RURAL MIGRATION
BT1 geographical mobility
BT2 migration

URBAN SOCIETY
BT1 societies

URBANIZATION
BT1 rural-urban relations
BT2 environment

UREA
BT1 nitrogen
BT2 inorganic fertilizers
BT3 fertilizers

URUGUAY
BT1 south america
BT2 america
rt lafta

USA
uf *united states of america*
BT1 north america
BT2 america
rt pacific islands us
rt puerto rico

USSR
uf *russia*
uf *soviet union*
uf *union of soviet socialist republics*
BT1 eastern europe
BT2 europe
rt byelorussian ssr
rt cmea
rt ukrainian ssr

USUFRUCT
BT1 ownership

UTILITY FUNCTIONS
BT1 single equation analysis
BT2 parametric methods
BT3 econometrics

UTILIZATION OF CAPACITY
BT1 efficiency
rt production possibilities

VALUATION
NT1 ground rent
NT1 land prices
NT1 rent
NT1 taxable rent value
rt compensation

VALUE ADDED
BT1 aggregate data

VALUE ADDED TAX
uf *vat*
BT1 indirect taxation
BT2 fiscal policy
BT3 public finance

VALUE SYSTEMS
BT1 culture

VALUES
BT1 culture

VARIABLE COSTS
BT1 production costs
BT2 costs
BT3 cost analysis

VARIABLE LEVIES
BT1 import levies

VARIABLE LEVIES (C)
BT2 import controls
BT2 levies
rt tariffs

VARIANCE
BT1 dispersion and distribution
BT2 descriptive statistics
BT3 statistics

VARIANCE ANALYSIS
BT1 statistical inference
BT2 statistical analysis
BT3 statistics

VARIETIES
uf *cultivars*
BT1 plant breeding
NT1 high yielding varieties

vat
USE **value added tax**

VATICAN
uf *holy see*
BT1 western europe
BT2 europe

VEAL
BT1 meat

VEGETABLE GROWING
BT1 horticulture
BT2 crop enterprises
BT3 farm enterprises
rt vegetables

VEGETABLE OILS
BT1 oils and fats
NT1 coconut oils
NT1 corn oils
NT1 cottonseed oils
NT1 groundnut oils
NT1 olive oil
NT1 palm oil
NT1 soy oils
NT1 sunflower seed oils

VEGETABLE PRODUCTS
NT1 potato products
NT1 soy products
rt food
rt prepared foods
rt vegetables

VEGETABLE PROTEINS
BT1 proteins

VEGETABLES
NT1 artichokes
NT1 asparagus
NT1 aubergines
NT1 beet
NT1 broccoli
NT1 brussels sprouts
NT1 cabbages
NT2 green cabbages
NT2 red cabbages
NT2 savoy cabbages
NT2 white cabbages
NT1 carrots
NT1 cauliflowers
NT1 celeriac
NT1 celery
NT1 chick peas
NT1 chicory
NT1 cucumbers
NT1 edible fungi
NT1 endives
NT1 fresh vegetables
NT1 garlic
NT1 gherkins
NT1 kohlrabi
NT1 leeks
NT1 legumes
NT2 beans
NT3 broad beans
NT3 french beans

VEGETABLES (C)
NT3 horse beans
NT3 kidney beans
NT3 lima beans
NT3 white beans
NT2 lentils
NT2 peas
NT3 forage peas
NT1 lettuces
NT1 marrows
NT1 okra
NT1 onions
NT1 parsnips
NT1 peppers
NT1 pumpkins
NT1 radishes
NT1 shallots
NT1 spinach
NT1 squashes
NT1 swedes
NT1 sweet corn
NT1 sweet potatoes
NT1 tomatoes
NT1 watercress
rt vegetable growing
rt vegetable products

VENEZUELA
BT1 south america
BT2 america
rt lafta

VERTICAL INTEGRATION
BT1 integration
BT2 socioeconomic organization
rt agribusiness
rt contract farming

VERTICAL MOBILITY
BT1 social mobility
BT2 social stratification
BT3 social structure
NT1 downward mobility
NT1 upward mobility

VETERINARY PRODUCTS
rt veterinary services

VETERINARY SERVICES
BT1 professional services
BT2 tertiary sector
BT3 economic sectors
rt animal diseases
rt veterinary products

VIETNAM
uf *north vietnam*
uf *south vietnam*
uf *vietnam dr*
uf *vietnam r*
BT1 indochina
BT2 south east asia
BT3 asia

vietnam dr
USE **vietnam**

vietnam r
USE **vietnam**

VILLAGES
BT1 rural communities
BT2 rural environment
BT3 environment
rt local authority areas

VINEYARDS
BT1 land
rt viticulture

virgin islands us
USE **united states virgin islands**

VIRGIN LANDS
BT1 land

VIRUS KILLERS
BT1 pesticides

VITAMINS
BT1 additives

VITICULTURE
BT1 crop enterprises
BT2 farm enterprises
rt grapes
rt vineyards

VITICULTURE CO-OPERATIVES
BT1 co-operatives

VOCATIONAL TRAINING
uf training
BT1 education
NT1 apprenticeship
NT1 on farm training
NT1 practical education
NT1 retraining

WAGE AGREEMENTS
BT1 labour economics

WAGE RATES
BT1 remuneration
BT2 labour economics

WAGES
uf salaries
sf labour payment
sf payment schemes
BT1 remuneration
BT2 labour economics

WAGES IN KIND
BT1 remuneration
BT2 labour economics

waiting line
USE **queueing theory**

WALNUTS
BT1 nuts

WARE POTATOES
BT1 potatoes
BT2 root crops
BT3 crops

WAREHOUSES
BT1 storage

WASTE DISPOSAL
uf effluent disposal
BT1 pollution by agriculture
BT2 environmental policy
rt agricultural wastes
rt slurry removal

WASTE LAND
BT1 land

WATER
rt irrigation
rt water management
rt water pollution
rt waterways

water buffaloes
USE **buffaloes**

WATER MANAGEMENT
BT1 land use systems
rt water
rt waterways

WATER POLLUTION
BT1 pollution of agriculture
BT2 environmental policy
rt water

WATERCRESS
BT1 vegetables

WATERMELONS
BT1 fruits

WATERSHEDS
uf catchment areas
uf river basins
BT1 land

WATERWAYS
rt water
rt water management

wealth tax
USE **capital taxation**

weaner pigs
USE **piglets**

weather
USE **climate**

WEED CONTROL
BT1 plant husbandry
BT2 cultivation methods

WEEDING
BT1 plant protection

WELFARE ECONOMICS
BT1 mixed economies
BT2 economic systems

WEST AFRICA
BT1 africa south of sahara
BT2 africa
NT1 benin
NT1 cameroon
NT1 cape verde
NT1 gambia
NT1 ghana
NT1 guinea
NT1 guinea-bissau
NT1 ivory coast
NT1 liberia
NT1 mali
NT1 mauritania
NT1 niger
NT1 nigeria
NT1 senegal
NT1 sierra leone
NT1 st helena
NT1 togo
NT1 upper volta
NT1 western sahara

WEST ASIA
uf middle east
uf near east
BT1 asia
NT1 afghanistan
NT1 cyprus
NT1 iran
NT1 iraq
NT1 israel
NT1 jordan
NT1 lebanon
NT1 oman
NT1 saudi arabia
NT1 syria
NT1 yemen
NT1 yemen pdr

west indies
USE **caribbean**

WEST INDIES ASSOCIATED STATES
BT1 caribbean
BT2 america
NT1 antigua
NT1 dominica
NT1 grenada
NT1 st kitts and nevis and anguilla
NT1 st lucia
NT1 st vincent
rt united kingdom

WESTERN EUROPE
BT1 europe
NT1 andorra
NT1 austria
NT1 azores

WESTERN EUROPE (C)
NT1 belgium
NT2 belgium (antwerp)
NT2 belgium (brabant)
NT2 belgium (east flanders)
NT2 belgium (hainault)
NT2 belgium (liege)
NT2 belgium (limbourg)
NT2 belgium (namur)
NT2 belgium (west flanders)
NT1 canary islands
NT1 cyprus
NT1 france
NT2 france (overseas depts and terr)
NT2 france (r. alsace)
NT2 france (r. aquitaine)
NT2 france (r. auvergne)
NT2 france (r. brittany)
NT2 france (r. burgundy)
NT2 france (r. centre)
NT2 france (r. champagne ardennes)
NT2 france (r. corsica)
NT2 france (r. franche comte)
NT2 france (r. ile de france)
NT2 france (r. languedoc roussillon)
NT2 france (r. limousin)
NT2 france (r. lorraine)
NT2 france (r. lower normandy)
NT2 france (r. midi pyrenees)
NT2 france (r. nord pas de calais)
NT2 france (r. pays de la loire)
NT2 france (r. picardy)
NT2 france (r. poitou charentes)
NT2 france (r. provence alpes cote d'az
NT2 france (r. rhone alpes)
NT2 france (r. upper normandy)
NT1 germany fr
NT2 germany fr (baden-wuerttemberg)
NT2 germany fr (bavaria)
NT2 germany fr (hesse)
NT2 germany fr (lower saxony)
NT2 germany fr (north rhine westphalia)
NT2 germany fr (rhineland palatinate)
NT2 germany fr (saarland)
NT2 germany fr (schleswig-holstein)
NT1 gibraltar
NT1 greece
NT1 ireland
NT2 ireland (connacht)
NT2 ireland (leinster)
NT2 ireland (munster)
NT1 italy
NT2 italy (abruzzi)
NT2 italy (apulia)
NT2 italy (basilicata)
NT2 italy (calabria)
NT2 italy (campania)
NT2 italy (emilia-romagna)
NT2 italy (friuli-venezia giulia)
NT2 italy (latium)
NT2 italy (liguria)
NT2 italy (lombardy)
NT2 italy (marches)
NT2 italy (molise)
NT2 italy (piedmont)
NT2 italy (sardinia)
NT2 italy (sicily)
NT2 italy (trentino-alto adige)
NT2 italy (tuscany)
NT2 italy (umbria)
NT2 italy (valle d'aosta)
NT2 italy (veneto)
NT1 liechtenstein
NT1 luxembourg
NT1 madeira
NT1 malta
NT1 monaco
NT1 netherlands
NT2 netherlands (drenthe)
NT2 netherlands (friesland)
NT2 netherlands (gelderland)
NT2 netherlands (groningen)
NT2 netherlands (n.brabant)
NT2 netherlands (n.holland)
NT2 netherlands (overijssel)
NT2 netherlands (s.holland)
NT2 netherlands (s.limburg)
NT2 netherlands (utrecht)

WESTERN EUROPE (C)
NT2 netherlands (zeeland)
NT1 portugal
NT1 san marino
NT1 scandinavia
NT2 denmark
NT3 denmark (bornholm)
NT3 denmark (fyn)
NT3 denmark (jutland)
NT3 denmark (lolland-falster)
NT3 denmark (zealand)
NT2 faroe islands
NT2 finland
NT2 greenland
NT2 iceland
NT2 norway
NT2 sweden
NT1 spain
NT1 switzerland
NT1 turkey
NT1 united kingdom
NT2 united kingdom (channel islands)
NT2 united kingdom (england and wales)
NT3 united kingdom (e.england)
NT3 united kingdom (e.midlands)
NT3 united kingdom (n.england)
NT3 united kingdom (s.w.england)
NT3 united kingdom (w.midlands)
NT3 united kingdom (wales)
NT3 united kingdom (yorks/lancs)
NT2 united kingdom (isle of man)
NT2 united kingdom (n.ireland)
NT2 united kingdom (scotland)
NT3 united kingdom (e.scotland)
NT3 united kingdom (highlands+islands)
NT3 united kingdom (n.scotland)
NT3 united kingdom (w.scotland)
NT1 vatican

western germany
USE **germany fr**

WESTERN SAHARA
uf *rio de oro*
uf *spanish sahara*
BT1 west africa
BT2 africa south of sahara
BT3 africa

WESTERN SAMOA
BT1 oceania
rt commonwealth

WHEAT
BT1 cereals
BT2 crops
NT1 hard wheat
NT1 soft wheat

WHEAT FLOUR
BT1 flour
BT2 cereal products

WHEY
BT1 milk products

WHITE BEANS
BT1 beans
BT2 legumes
BT3 vegetables

WHITE CABBAGES
BT1 cabbages
BT2 vegetables

WHO
(world health organization)
BT1 organizations

whole milk powder
USE **dried milk**

WHOLESALE MARKETING
BT1 marketing channels
BT2 marketing
NT1 buying groups
NT1 product markets

WHOLESALE PRICES
BT1 prices

WILD FRUITS
BT1 fruits

WILD LIFE
NT1 natural fauna
NT1 natural flora

WINE
BT1 alcoholic beverages
BT2 beverages
NT1 champagne and sparkling wines
NT1 quality wine
NT1 table wine
rt grapes

WINE INDUSTRY
BT1 beverage industry
BT2 food industry

WITHDRAWAL PRICES
BT1 prices

WOMAN'S STATUS
BT1 family life
rt women

WOMEN
uf *housewives+*
NT1 rural women
rt female labour
rt woman's status

WOOD INDUSTRY
BT1 non-food industries

WOOL INDUSTRY
BT1 textile industries
BT2 non-food industries

WOOL SHEEP BREEDS
BT1 sheep

work
USE **labour**

WORK BANKS
uf *labour banks*
BT1 farm helper services
BT2 co-operative services

WORK CONFLICTS
BT1 working conditions
rt social crisis

WORK NORMS
BT1 work planning
BT2 labour economics

WORK PLACES
BT1 working conditions

WORK PLANNING
BT1 labour economics
NT1 labour contracts
NT1 labour requirements
NT1 work norms
rt planning

WORK SATISFACTION
uf *job satisfaction*
BT1 working conditions

WORK STRESS
BT1 working conditions

WORK STUDY
uf *time study*
BT1 labour economics
NT1 rest periods
NT1 unoccupied time
NT1 working hours
rt sociology of work

WORKERS
BT1 occupations

WORKING CAPITAL
uf *circulating funds*
BT1 assets
BT2 balance sheets
BT3 balance sheet analysis
NT1 liquid assets
rt capital

WORKING CLASSES
uf *proletariat*
BT1 social classes
BT2 social stratification
BT3 social structure

WORKING CONDITIONS
NT1 attitudes to work
NT1 retirement
NT1 safety at work
NT2 accident prevention
NT2 industrial illness
NT1 work conflicts
NT1 work places
NT1 work satisfaction
NT1 work stress
NT1 working hours
rt labour economics
rt labour protection
rt sociology of work

WORKING GROUPS
BT1 groups

WORKING HOURS
BT1 work study
BT2 labour economics
BT1 working conditions

WORKING POPULATION
BT1 population structure
BT2 population
NT1 agricultural work force
rt agricultural manpower

WORLD BANK
(international bank for reconstruction and development)
uf *ibrd*
BT1 organizations
rt investment banks

WORLD MARKETS
BT1 markets

world trade
USE **international trade**

YAMS
BT1 root crops
BT2 crops

YEMEN
uf *yemen arab republic*
BT1 arab countries
BT1 west asia
BT2 asia

yemen arab republic
USE **yemen**

YEMEN PDR
uf *aden*
uf *yemen south*
BT1 arab countries
BT1 west asia
BT2 asia

yemen south
USE **yemen pdr**

YIELDS
rt high yielding varieties
rt productivity

YOGHURT
BT1 cultured milks
BT2 milk products

YOUNG WORKERS
BT1 agricultural manpower

YOUTH
BT1 age
BT1 family size
 BT2 families
NT1 rural youth

YUGOSLAVIA
uf *jugoslavia*
BT1 eastern europe
 BT2 europe
BT1 mediterranean countries

ZAIRE
uf *congo (kinshasa)*
uf *congo (leopoldville)*
uf *congo democratic republic*
BT1 central africa
 BT2 africa south of sahara
  BT3 africa
rt french-speaking africa

ZAMBIA
BT1 southern africa
 BT2 africa south of sahara
  BT3 africa
rt commonwealth
rt english-speaking africa

zones
USE **agricultural regions**

ZONING
BT1 physical planning
 BT2 planning
rt building controls
rt land use

MICROTHESAURI

ACQUISITION OF OWNERSHIP
NT1 accession
NT1 contracts
NT1 possession
NT1 property protection
NT1 succession
NT2 indemnification
NT2 indivisible estates
NT2 partition of estates

ADMINISTRATION
NT1 central government
NT2 ministries of agriculture
NT1 local government
NT2 local authority areas
NT1 private organizations
NT1 public authorities
NT1 regional government
NT1 semi-private organizations

AGRICULTURAL LAW
rt contract legislation

AGRICULTURAL TENURE
NT1 farm leases
NT1 grazing tenancy
NT1 heritable leases
NT1 livestock leases
rt landowners
rt leases
rt period of tenancy
rt resumption of possession
rt tenants

AGRICULTURAL TRIBUNALS
rt jurisprudence

CADASTERS
rt land ownership

CASE LAW
NT1 jurisprudence
NT1 legal principles
NT1 practice

COMMERCIAL LAW
rt company law
rt legal form
rt trade

COMPANY LAW
rt commercial law
rt companies
rt legal form

CONTRACT LEGISLATION
rt agricultural law
rt integration
rt social legislation

EC LEGAL DECISIONS
rt practice
rt treaty of rome

FISHING
rt farm enterprises
rt fish farming
rt leisure

FOREST OWNERSHIP
NT1 national forests
NT1 private forests
NT1 public forests

LAND OWNERSHIP
NT1 property transfers
NT2 inheritance
NT2 land transfers
NT2 mortgages
rt cadasters
rt ownership
rt ownership conditions

LAND USE
rt cultivation methods
rt land
rt landscape conservation
rt zoning

LAND USE SYSTEMS
NT1 forest management
NT2 afforestation
NT1 public domain
NT2 common lands
NT2 national forests
NT2 national parks
NT1 water management
rt forest policy

LANDLORDS RIGHT
rt landowners

LANDOWNERS
rt agricultural tenure
rt landlords right
rt resumption of possession
rt tenants

LEGAL FORM
rt commercial law
rt companies
rt company law
rt legislation

LEGAL SOURCES
NT1 collective agreements
NT1 constitutional law
NT1 customary law
NT1 ec laws
NT2 treaty of rome
NT3 ec regulations
NT1 legislation
NT2 civil law
NT2 code forestier
NT2 codified rural legislation
NT2 laws
NT3 subnational legislation
NT2 regulations
NT3 decrees
NT1 local byelaws
NT1 professional codes

LICENSES AND PERMITS
NT1 building permits
rt regulations

NON-TRANSFERABLE PROPERTY
rt indivisible estates
rt land mobility

OPTION TO PURCHASE
rt contracts
rt termination of leases

OWNERSHIP
NT1 co-ownership
NT1 ownership limitation
NT2 adverse possession
NT2 compulsory purchase
NT2 easements
NT1 private ownership
NT1 public ownership
NT1 usufruct
rt land consolidation
rt land ownership

PERIOD OF TENANCY
rt agricultural tenure

PUBLIC HEALTH LEGISLATION
NT1 food inspection
rt animal diseases
rt plant diseases

ROADS
NT1 farm roads
NT1 rural roads

SOCIAL LEGISLATION
NT1 labour legislation
NT2 labour protection
NT2 strikes
rt contract legislation

SOCIAL POLICY
NT1 rural social policy
rt legislation
rt policy
rt social security

TENANTS
rt agricultural tenure
rt landowners
rt tenants rights

TENANTS RIGHTS
rt tenants

TENURE SYSTEMS
NT1 metayage
NT1 owner occupancy
NT1 tenancy
NT2 leases

TERMINATION OF LEASES
NT1 resumption of possession
rt option to purchase

WATERWAYS
rt water
rt water management

ACCOUNTING
NT1 aggregate farm accounts
rt aggregate accounts
rt balance sheet analysis
rt book-keeping
rt cost analysis
rt farm planning
rt farm results
rt income
rt input-output analysis
rt national accounting
rt regional accounting

AGGREGATE ACCOUNTS
NT1 national accounting
NT1 regional accounting
rt accounting

AGGREGATE DATA
NT1 consumer expenditure
NT2 household expenditure
NT1 final consumption
NT1 final production
NT1 gross national product
NT1 household consumption
NT1 intermediate consumption
NT1 national expenditure
NT1 total agricultural expenditure
NT1 value added
rt economic accounts

AGRARIAN REFORM
NT1 land reform
rt economic development
rt structural policy

AGRICULTURAL AVIATION
NT1 aerial surveys
rt air transport

AGRICULTURAL BILLS OF EXCHANGE
rt credit transactions

AGRICULTURAL DEVELOPMENT
NT1 agricultural crisis
NT1 green revolution
NT1 rural-urban migration
rt agricultural situation
rt sectoral development

AGRICULTURAL DISASTERS
NT1 fires
NT2 forest fires
NT1 flooding
NT1 tempest
rt climatic factors
rt compensation
rt crop insurance
rt damage
rt natural disasters
rt subsidies

AGRICULTURAL ENGINEERING
rt land improvement
rt mechanization
rt professional services
rt technical progress

AGRICULTURAL FINANCIAL POLICY
NT1 subsidies
NT2 acreage subsidies
NT2 grubbing up grants
NT2 improvement grants
NT2 non-marketing premiums
NT2 setting-up grants
NT2 severance allowance
NT2 slaughter premiums
NT2 stock-piling subsidies
rt agricultural policy
rt policy

AGRICULTURAL MANPOWER
NT1 child labour
NT1 elderly family-workers

AGRICULTURAL MANPOWER (C)
NT1 family labour
NT1 hired labour
NT2 female labour
NT2 migrant labour
NT2 part time labour
NT2 permanent labour
NT2 seasonal labour
NT2 skilled labour
NT2 unskilled labour
NT1 manpower units
NT1 young workers
rt agricultural work force
rt farm workers
rt labour
rt labour economics
rt working population

AGRICULTURAL POLICY
NT1 cap
NT1 marketing policy
NT2 market regulations
NT3 intervention
NT2 market stabilization
NT2 price policy
NT3 compensatory amounts
NT1 national agricultural policy
NT1 regional agricultural policy
NT2 less favoured areas
NT3 mountain areas
NT1 structural policy
NT2 land consolidation
NT1 support measures
rt agricultural financial policy
rt agricultural trade
rt food policy
rt forest policy
rt land diversion
rt policy
rt production policy

AGRICULTURAL SITUATION
NT1 capital allocation
NT1 disparity
NT1 farm indebtedness
NT2 bankruptcy
NT1 investment
NT2 foreign investment
NT2 private investment
NT2 public investment
NT2 reinvestment
NT1 parity
NT1 productivity
NT2 capital productivity
NT2 labour productivity
NT2 land productivity
NT1 resource allocation
rt agricultural development
rt economic analysis

AGRICULTURAL STRUCTURE
NT1 farm structure
NT1 production structure
rt structure

AGRICULTURAL SYSTEMS
NT1 commercial farming
NT1 extensive farming
NT1 intensive farming
NT1 subsistence farming
NT1 transitional farming
NT1 types of farming
NT2 collective farming
NT2 domestic gardens
NT2 family farms
NT2 full-time farming
NT2 large farms
NT2 medium sized farms
NT2 mixed farming
NT2 part-time farming
NT3 peasant workers
NT2 plantations
NT2 small farms
NT2 specialized farming

AGRICULTURAL SYSTEMS (C)
rt farm structure

AGRO-INDUSTRIAL SECTOR
NT1 agribusiness
NT1 post-agricultural sector
NT1 pre-agricultural sector

ANIMAL DISEASES
rt parasites
rt public health legislation
rt veterinary services

BALANCE OF PAYMENTS
NT1 terms of trade
rt balance of trade
rt external debt

BALANCE SHEET ANALYSIS
NT1 accounting balance
NT1 annual accounts
NT1 balance sheets
NT2 assets
NT3 fixed capital
NT3 livestock capital
NT3 working capital
NT4 liquid assets
NT2 cash flow
NT2 liabilities
NT3 balance adjustment items
NT4 consolidated balance sheets
NT3 outside capital
NT3 owners equity
NT3 reserve funds
NT1 replacement value
rt accounting

BANKS
NT1 agricultural banks
NT1 co-operative banks
NT2 co-operative credit
NT1 commercial banks
NT1 development banks
NT1 industrial banks
NT1 investment banks
NT1 savings banks
rt financial institutions

BOOK-KEEPING
NT1 amortization
NT2 amortization methods
NT1 book-keeping systems
NT2 double entry book-keeping
NT2 global accounting
NT2 single entry book-keeping
NT1 inventories
NT1 plan comptable
NT1 stock accounting
rt accounting

CLIMATE
NT1 microclimate
rt climatic areas
rt meteorology

CLIMATIC AREAS
NT1 arid zones
NT2 deserts
NT3 desertification
NT1 humid zones
NT1 temperate zones
NT1 tropics
rt climate

CLIMATIC FACTORS
NT1 drought
NT1 frost
NT1 precipitation
NT2 hail
NT2 rainfall
NT2 snow
rt agricultural disasters
rt damage

CO-OPERATIVE ACTIVITIES
NT1 co-operative bargaining
NT1 co-operative farming
NT1 co-operative marketing
NT2 co-operative purchasing
NT2 co-operative selling
NT1 co-operative processing
rt co-operatives

CO-OPERATIVE FARM ENTERPRISES
(cooperation on a single crop or livestock activity)
NT1 crop production groups
NT1 forestry groups
NT1 gaec
NT1 grazing groups
NT1 groupement foncier agricole
NT1 livestock groups
NT1 producer groups
NT1 safer
rt co-operation
rt co-operative farming
rt collective farming

CO-OPERATIVE SERVICES
NT1 breeders' improvement societies
NT1 co-operative banks
NT2 co-operative credit
NT1 co-operative insurance
NT1 co-operative plant protection
NT1 farm helper services
NT2 co-operative feedstores
NT2 machinery co-operatives
NT2 work banks
NT1 land improvment co-operatives
rt co-operation

CO-OPERATIVES
NT1 consumer co-operatives
NT1 dairy co-operatives
NT1 multi-purpose co-operatives
NT1 mutual organisations
NT1 producer co-operatives
NT1 viticulture co-operatives
rt co-operation
rt co-operative activities
rt private firms

COMPENSATION
rt agricultural disasters
rt compulsory purchase
rt insurance benefits
rt valuation

CONSUMER PROTECTION
NT1 consumer information
NT1 prevention of fraud
NT1 quality controls
NT2 designation of origin
NT2 labelling controls
rt food inspection

CONSUMPTION
NT1 consumption per caput
NT1 consumption units
NT1 domestic consumption
NT1 household consumption
NT1 on farm consumption
rt demand

CONTRACT FARMING
rt vertical integration

COST ANALYSIS
NT1 costing
NT1 costs
NT2 production costs
NT3 actual costs
NT3 estimated costs
NT3 fixed costs
NT3 special costs
NT3 total costs
NT3 variable costs
NT1 returns

COST ANALYSIS (C)
rt accounting

CREDIT POLICY
NT1 credit control
NT2 discount rates
NT2 interest rates
NT3 preferential interest rates
NT1 repayment
NT2 debt consolidation
NT2 interest repayments
rt credit transactions
rt loans
rt policy

CREDIT TRANSACTIONS
NT1 credit
NT2 agricultural credit
NT2 commercial credit
NT3 capital leasing
NT2 export credits
NT2 long term credit
NT2 medium term credit
NT2 short term credit
NT1 debt
NT2 agricultural debentures
NT2 external debt
NT2 public debt
NT1 hire purchase
NT1 loans
NT2 international loans
NT2 private loans
NT3 bank loans
NT3 mortgages
NT2 public loans
rt agricultural bills of exchange
rt credit policy

CROPPING SYSTEMS
NT1 dry farming
NT1 greenhouse crops
NT1 hydroponics
NT1 irrigated farming
NT1 organic farming
NT1 perennial cropping
NT1 primitive rotation farming
NT1 terracing
rt crops

CULTIVATION METHODS
NT1 harvesting
NT1 land clearance
NT1 land improvement
NT2 drainage
NT2 fertilizer application
NT2 irrigation
NT2 soil conservation
NT2 spray irrigation
NT1 plant husbandry
NT2 biological control
NT2 grafting
NT2 integrated plant protection
NT2 pest control
NT2 pruning
NT2 thinning
NT2 weed control
NT1 rotation
NT1 tillage
NT2 harrowing
NT2 minimal tillage systems
NT2 no-tillage system
NT2 planting
NT2 ploughing
NT2 replanting
NT2 rolling
NT2 sowing
NT2 subsoiling
rt land use

CURRENCY
NT1 blocked currencies
NT1 european units of account
NT1 foreign exchange
NT1 green money

CURRENCY (C)
NT1 national currencies

DAMAGE
rt agricultural disasters
rt climatic factors
rt subsidies

DEVELOPMENT AID
NT1 food aid
NT1 technical aid
rt developing countries
rt development policy
rt development projects
rt funds
rt international agreements

DEVELOPMENT POLICY
rt development aid
rt policy

ECONOMIC ACCOUNTS
NT1 capital accounts
NT1 farm accounts
NT1 income accounts
NT1 production accounts
rt aggregate data

ECONOMIC ANALYSIS
NT1 marginal analysis
NT2 marginal costs
NT2 marginal returns
rt agricultural situation
rt gross margins analysis
rt input-output analysis
rt microeconomic analysis

ECONOMIC DEVELOPMENT
NT1 economic growth
NT2 growth rate
NT1 regional development
NT1 sectoral development
NT1 trade cycles
NT2 economic depression
NT2 economic recovery
NT2 fluctuations
NT3 cyclic fluctuations
NT3 seasonal fluctuations
NT3 trends
NT2 stabilization
NT2 stagnation
rt agrarian reform
rt social change

ECONOMIC GEOGRAPHY
NT1 agrarian countries
NT1 developed countries
NT1 developing countries
NT1 least developed countries
rt geography

ECONOMIC SECTORS
NT1 primary sector
NT2 farm sector
NT1 secondary sector
NT1 tertiary sector
NT2 professional services
NT3 veterinary services
NT2 public services
rt sectoral development

ECONOMIC SITUATION
NT1 economic resources
NT2 capital
NT2 energy
NT3 electricity
NT3 solar energy
NT2 human resources
NT3 labour
NT3 management
NT4 farm management
NT4 self management
NT2 natural resources
NT2 production possibilities

ECONOMIC SOCIOLOGY
NT1 consumer behaviour
NT2 eating habits
NT2 purchasing habits
NT1 farmers attitudes
rt consumer expenditure
rt consumer motivation

ECONOMIC SYSTEMS
NT1 centrally planned economies
NT1 market economies
NT2 competitive economies
NT1 mixed economies
NT2 welfare economics

ECONOMIC THEORIES
NT1 development theory
NT2 economic dualism
NT1 economic laws
NT2 economies of scale
NT1 economic schools
NT2 classical economics
NT2 keynsian economics
NT2 marxist economics
NT2 neo-classical economics
NT1 equilibrium theory
NT1 growth theory
NT1 location theory
rt economics
rt market economics

ECONOMIC UNIONS
NT1 common markets
NT1 customs unions
rt monetary unions

ECONOMIC UNITS
NT1 financial institutions
NT1 households
NT2 agricultural households
NT1 production unit
rt manpower units

ECONOMICS
NT1 agricultural economics
rt economic theories

EFFICIENCY
NT1 utilization of capacity
rt evaluation
rt profit and loss analysis

ELECTRIFICATION
rt electricity
rt farm equipment

ENVIRONMENTAL POLICY
NT1 landscape conservation
NT1 nature conservancy
NT1 pollution by agriculture
NT2 agricultural wastes
NT3 crop residues
NT3 livestock wastes
NT2 waste disposal
NT1 pollution control
NT1 pollution of agriculture
NT2 air pollution
NT2 water pollution
rt environment
rt forest policy
rt physical planning
rt policy

FARM AREA
NT1 arable land
NT1 farmland
rt farm size

FARM BUILDINGS
NT1 livestock housing
rt farm inputs
rt fixed capital

FARM CLOSURES
rt severance allowance
rt structural policy

FARM ENTERPRISES
(branches of farming)
NT1 ancillary enterprises
NT1 crop enterprises
NT2 field crops
NT2 forestry
NT2 grazing
NT2 horticulture
NT3 floriculture
NT3 fruit growing
NT3 tree nurseries
NT3 vegetable growing
NT2 ley farming
NT2 viticulture
NT1 food production
NT1 livestock enterprises
NT2 apiculture
NT2 aquaculture
NT2 cattle farming
NT3 dairy farming
NT2 fish farming
NT2 goat keeping
NT2 horse breeding
NT2 pig farming
NT2 poultry farming
NT2 sheep farming
NT2 silk worms
NT2 small animal rearing
NT1 mixed cropping
rt farms
rt fishing
rt hunting

FARM EQUIPMENT
NT1 farm machinery
NT2 harvester
NT3 combine harvester
NT2 tractors
NT1 tools
rt electrification
rt farm inputs
rt fixed capital
rt mechanization

FARM INPUTS
rt farm buildings
rt farm equipment
rt feedingstuffs
rt fertilizers
rt fuels
rt pesticides
rt plants
rt seeds
rt selective herbicides

FARM PLANNING
NT1 farm budgeting
NT1 farm models
NT1 financial planning
NT2 budget accounting
NT1 gross margins analysis
NT1 investment planning
NT2 risk
NT2 uncertainty
NT1 management by objectives
rt accounting
rt agricultural planning
rt farm plans
rt planning

FARM RESULTS
NT1 earned income
NT1 farm income
NT1 farm indebtedness
NT2 bankruptcy
NT1 farm profit
NT1 farmers income
NT2 non-farm income
NT1 gross returns
NT1 losses

FARM RESULTS (C)
NT1 net returns
rt accounting

FARM SIZE
NT1 livestock numbers
NT1 parcel size
rt farm area
rt farm structure
rt large farms
rt medium sized farms
rt small farms

FARMS
NT1 collective farms
NT2 kibbutzim
NT2 kolkhozy
NT2 moshavim
NT1 private farms
NT1 state farms
rt farm enterprises
rt types of farming

FEEDINGSTUFFS
NT1 farm produced fodder
NT2 dry feed
NT2 green fodder
NT2 hay
NT2 roughage
NT2 silage
NT2 straw
NT1 manufactured feeds
NT2 concentrates
NT2 feed cake
NT2 feed-mixes
rt additives
rt farm inputs
rt feed crops
rt feeding

FERTILIZERS
NT1 inorganic fertilizers
NT2 compost
NT2 compound fertilizers
NT2 lime
NT2 nitrogen
NT3 urea
NT2 phosphates
NT2 potash
NT1 organic fertilizers
NT2 green manure
NT2 guano
NT2 manure
NT2 slurry
rt farm inputs

FINANCE
NT1 investment income
NT1 outside finance
NT1 savings
NT1 self finance

FIRMS
NT1 conglomerates
NT1 foreign firms
NT1 multinational corporations
NT1 nationalized enterprises
NT1 private firms
NT2 companies
NT2 private companies
NT1 public enterprises

FOOD INDUSTRY
NT1 bakery industry
NT2 pasta industry
NT1 beverage industry
NT2 brewing industry
NT2 coffee industry
NT2 distilling industry
NT2 tea industry
NT2 wine industry
NT1 canning industry
NT1 catering industry
NT1 cocoa industry

FOOD INDUSTRY (C)
NT1 confectionery industry
NT1 dairy industry
NT1 fish industry
NT1 frozen foods industry
NT1 meat and livestock industry
NT2 abattoirs
NT1 milling industry
NT1 oils and fats industry
NT1 starch industry
NT1 sugar industry
rt industry
rt post-agricultural sector
rt processing

FOOD POLICY
rt agricultural policy
rt food aid
rt food production
rt food supply
rt hunger
rt nutrition
rt policy

FOREST POLICY
rt agricultural policy
rt environmental policy
rt forest land
rt land use systems
rt policy

FUELS
rt electricity
rt energy
rt farm inputs

FUNDS
NT1 eaggf
NT1 european regional development fund
NT1 european social fund
NT1 national fund
rt development aid
rt investment promotion
rt marketing policy
rt regional agricultural policy
rt regional development

GEOGRAPHY
NT1 landscape
NT1 topography
rt agricultural geography
rt economic geography
rt maps
rt social geography

HANDLING
NT1 bulk handling
rt storage
rt transport

HUNTING
rt farm enterprises
rt game
rt leisure

IMPORT CONTROLS
NT1 import levies
NT2 variable levies
rt trade barriers

INCOME
NT1 income distribution
NT2 income transfers
rt accounting
rt national accounting

INDUSTRY
NT1 rural industry
rt food industry
rt input industries
rt non-food industries

INFRASTRUCTURE
rt rural planning

INPUT INDUSTRIES
NT1 agricultural machinery industry
NT1 chemical industry
NT2 fertilizer industry
NT1 feed industry
NT1 seed industry
rt industry
rt pre-agricultural sector
rt processing

INSURANCES
NT1 agricultural insurance
NT2 crop insurance
NT2 fire insurance
NT2 hail insurance
NT2 livestock insurance
NT1 credit insurance
NT1 export insurance
NT1 insurance benefits
NT1 insurance premiums
rt accident benefits
rt health services
rt maternity benefits
rt old age benefits
rt sickness benefits
rt social insurance
rt social security
rt social services
rt unemployment insurance

INVESTMENT POLICY
NT1 investment promotion
NT2 farm plans
NT3 investment support thresholds
NT2 investment guarantees
NT1 investment requirements
rt policy

LABOUR ECONOMICS
NT1 division of labour
NT1 labour contracts
NT1 labour costs
NT1 organization of work
NT2 labour allocation
NT1 remuneration
NT2 piece work wages
NT2 time wages
NT2 wage rates
NT2 wages
NT2 wages in kind
NT1 wage agreements
NT1 work planning
NT2 labour contracts
NT2 labour requirements
NT2 work norms
NT1 work study
NT2 rest periods
NT2 unoccupied time
NT2 working hours
rt agricultural manpower
rt agricultural work force
rt employment
rt labour
rt labour legislation
rt leisure
rt working conditions

LAND
NT1 abandoned land
NT1 arable land
NT1 fallow
NT1 forest land
NT1 orchards
NT1 pastures
NT2 permanent pastures
NT2 prairies
NT2 temporary grasslands
NT1 reclaimed land
NT2 polder
NT1 vineyards
NT1 virgin lands
NT1 waste land
NT1 watersheds
rt factors of production

LAND (C)
rt land use
rt natural resources

LAND BANKS
rt land diversion
rt safer

LEVIES
NT1 import levies
NT2 variable levies
rt fiscal policy

LICENSES AND PERMITS
NT1 building permits
rt regulations

LIVESTOCK SECTOR
NT1 animal breeding
NT2 artificial insemination
NT2 breeds
NT2 cross breeding
NT2 mating
NT2 selective breeding
NT1 animal welfare
NT1 feeding
NT2 fattening
NT3 extensive production
NT3 indoor fattening
NT3 intensive fattening
NT2 feed rations
NT2 feed requirements
NT2 fodder conversion
NT1 herd improvement
NT1 livestock farming
NT2 extensive livestock farming
NT2 hand rearing
NT2 industrial rearing
NT1 livestock husbandry
NT2 controlled grazing
NT2 grazing
NT2 slurry removal
NT2 stabling
NT2 transhumance
NT1 milking

MARKET ECONOMICS
NT1 competition
NT2 imperfect competition
NT3 monopoly
NT3 monopsony
NT3 oligopoly
NT3 oligopsony
NT2 perfect competition
NT1 demand
NT1 market transparency
NT2 market intelligence
NT1 supply
NT1 supply and demand
rt economic theories
rt market economies

MARKET RESEARCH
NT1 consumer surveys
NT2 consumer motivation
NT2 consumer panels
NT1 market surveys
rt marketing
rt marketing channels

MARKETING
NT1 distribution
NT1 marketing channels
NT2 commodity exchanges
NT3 futures markets
NT2 merchants
NT3 comission agents
NT3 middlemen
NT2 retail marketing
NT3 direct marketing
NT3 street markets
NT3 supermarkets
NT2 wholesale marketing
NT3 buying groups

MARKETING (C)
NT3 product markets
rt market research

MARKETING MARGINS
rt distribution
rt marketing channels

MARKETING TECHNIQUES
NT1 auctions
NT1 bonuses
NT1 credit sales
NT1 discounts
NT1 door to door sales
NT1 mail order selling
NT1 market segmentation
NT1 sales promotion
NT2 labelling
NT3 quality labelling
NT4 standard labelling
NT3 trade marks
NT2 launching a product
NT2 product image
NT2 product presentation
NT2 publicity
NT3 advertizing
NT3 trade fairs
NT2 special offers
rt purchasing
rt sales

MARKETS
NT1 capital market
NT1 commodity markets
NT1 domestic markets
NT1 intra-community markets
NT1 labour market
NT2 labour mobility
NT1 land markets
NT2 land mobility
NT1 world markets

MECHANIZATION
NT1 maintenance
rt agricultural engineering
rt automation
rt farm equipment

METEOROLOGY
rt climate

MICROECONOMIC ANALYSIS
NT1 factors of production
rt economic analysis

MONETARY POLICY
NT1 monetary agreements
NT1 monetary parity
NT2 devaluation
NT2 revalution
NT1 monetary reform
NT1 monetary unions
rt policy

MONETARY SITUATION
NT1 deflation
NT1 inflation

MONETARY SYSTEMS
NT1 drawing rights

NON-FOOD INDUSTRIES
NT1 leather industry
NT1 paper industry
NT1 rubber industry
NT1 textile industries
NT2 cotton industry
NT3 cotton ginning
NT2 wool industry
NT1 tobacco industry
NT1 wood industry
rt industry
rt processing

NON-TARIFF BARRIERS TO TRADE
NT1 tax agreements
rt trade barriers

NOXIOUS SUBSTANCES
rt pollution by agriculture
rt pollution of agriculture

NUTRITION
rt eating habits
rt food policy
rt food supply
rt health
rt proteins

OCCUPATIONS
NT1 farm workers
NT1 farmers
NT1 managers
NT1 teachers
NT1 workers

PACKAGING
NT1 bottling
NT1 canning
rt labelling
rt packs
rt preservation

PACKS
rt packaging
rt product presentation

PARASITES
rt animal diseases
rt pesticides
rt plant diseases

PESTICIDES
NT1 acaricides
NT1 bactericides
NT1 corvicides
NT1 fungicides
NT1 herbicides
NT1 insecticides
NT1 larvicides
NT1 mole-killer
NT1 nematicides
NT1 ovicides
NT1 rodenticides
NT1 selective herbicides
NT1 virus killers
rt farm inputs
rt parasites
rt phytotoxicity

PHYTOTOXICITY
rt pesticides

PLANT BREEDING
NT1 hybridization
NT1 plants
NT1 seeds
NT1 varieties
NT2 high yielding varieties

PLANT DISEASES
rt parasites
rt public health legislation

PLANT PROTECTION
NT1 burning
NT1 fumigation
NT1 spraying
NT1 weeding
rt integrated plant protection

POLICY
(use only for policies not specifically listed)
rt agricultural financial policy
rt agricultural policy
rt cap
rt credit policy
rt development policy

POLICY (C)
rt educational policy
rt environmental policy
rt food policy
rt forest policy
rt intervention
rt investment policy
rt marketing policy
rt monetary policy
rt national agricultural policy
rt price policy
rt production policy
rt regional agricultural policy
rt research policy
rt rural social policy
rt social policy
rt structural policy
rt support measures
rt trade policy

POLITICAL SYSTEMS
NT1 capitalism
NT1 communism
NT1 democracy
NT2 parliament
NT1 dictatorship
NT1 federalism
NT1 feudalism
NT1 imperialism
NT2 colonialism
NT1 socialism
rt nationalism

PRESERVATION
NT1 cooling
NT1 drying
NT1 freezing
rt additives
rt packaging
rt processing
rt silage

PRICE FORMATION
NT1 price controls
NT1 price elasticity
NT1 price fixing
NT1 price stabilization
NT1 price support

PRICES
NT1 activating prices
NT1 agricultural prices
NT2 guaranteed minimum producer prices
NT2 guide prices
NT2 intervention prices
NT1 average prices
NT1 basic prices
NT1 buying prices
NT1 consumer prices
NT1 controlled prices
NT1 cost prices
NT1 food prices
NT1 frozen prices
NT1 guide prices
NT1 input prices
NT1 market prices
NT1 maximum prices
NT1 minimum prices
NT1 norm prices
NT1 producer prices
NT1 reference prices
NT1 retail prices
NT1 selling prices
NT1 sluicegate prices
NT1 target prices
NT1 threshold prices
NT1 wholesale prices
NT1 withdrawal prices
rt price policy

PROCESSING
rt food industry
rt input industries
rt non-food industries

PROCESSING (C)
rt preservation
rt processed products

PRODUCTION
NT1 domestic production
rt production policy
rt supply

PRODUCTION POLICY
NT1 adjustment of production
NT2 agricultural adjustment
NT2 dairy-beef conversion schemes
NT1 land diversion
NT1 location of production
NT1 production controls
NT2 agricultural guidelines
NT2 agro-industrial relations
NT2 surpluses
rt agricultural policy
rt policy
rt production
rt structural policy

PROFIT AND LOSS ANALYSIS
NT1 liquidity
NT1 productivity
NT2 capital productivity
NT2 labour productivity
NT2 land productivity
NT1 profitability
NT2 break-even point
NT2 gross margins
rt efficiency

PROTECTION
NT1 discrimination
NT2 boycotts
NT2 dumping
NT2 embargos
NT2 price discrimination
NT1 quotas
NT1 trade barriers
NT2 export controls

PUBLIC FINANCE
NT1 agricultural budget
NT1 fiscal policy
NT2 direct taxation
NT3 capital taxation
NT4 land tax
NT3 company tax
NT3 income tax
NT3 inheritance tax
NT2 indirect taxation
NT3 exise duties
NT3 stamp duty
NT3 value added tax
NT2 taxes
rt public debt
rt public loans

QUALITY
rt quality labelling
rt quality standards

REGIONS
NT1 administrative areas
NT1 agricultural regions
NT1 economic regions

SOCIAL SECURITY
NT1 accident benefits
NT1 family allowances
NT1 health services
NT1 life insurance
NT1 maternity benefits
NT1 old age benefits
NT1 social insurance
NT2 invalid benefits
NT2 sickness benefits
NT1 social services
NT1 unemployment insurance
rt insurances

SOCIAL SECURITY (C)
rt legal systems
rt rural social policy
rt social policy

SOCIOECONOMIC ORGANIZATION
NT1 centralisation
NT1 co-operation
NT1 concentration
NT2 farm amalgamation
NT1 coordination
NT1 decentralization
NT1 diversification
NT1 integration
NT2 horizontal integration
NT2 vertical integration
NT1 rationalization
NT1 specialization

SOIL
rt soil analysis
rt soil properties

SOIL ANALYSIS
rt soil
rt soil properties

SOIL PROPERTIES
NT1 soil fertility
NT1 soil humidity
rt soil
rt soil analysis

STANDARDS
NT1 date stamping
NT1 grading
NT1 quality standards
NT1 quantity controls
rt social standards

STOCKS
NT1 buffer stocks
rt stock-piling subsidies
rt storage
rt supply balance
rt surpluses

STORAGE
NT1 bulk storage
NT1 cold storage
NT1 controlled atmosphere storage
NT1 silos
NT1 storage tanks
NT1 warehouses
rt handling
rt stocks

STRUCTURE
(use only for types of structure not separatly listed)
rt agricultural structure
rt family structure
rt farm structure
rt population structure
rt production structure
rt social structure

SUBSTITUTION
rt elasticities
rt simulated foods

SUPPLY BALANCE
NT1 food supply
NT1 self sufficiency
NT1 share of the market
rt balance of trade
rt demand
rt stocks
rt supply
rt supply and demand

TARIFFS
NT1 preferential tariffs
rt indirect taxation

TARIFFS (C)
rt trade barriers
rt variable levies

TENURE SYSTEMS
NT1 metayage
NT1 owner occupancy
NT1 tenancy
NT2 leases

TOURISM
NT1 farm holidays
rt ancillary enterprises

TRADE
NT1 barter
NT1 domestic trade
NT1 external trade
NT2 agricultural trade
NT2 exports
NT2 imports
NT1 international trade
NT1 intra-community trade
NT1 purchasing
NT1 sales
NT1 transit trade
NT1 turnover
rt agricultural trade
rt commercial law

TRADE POLICY
NT1 balance of trade
NT1 commercial credit
NT2 capital leasing
NT1 export diversification
NT1 export promotion
NT2 export credits
NT2 export refunds
NT2 export subsidies
NT1 imports promotion
NT1 marketing boards
NT1 trade agreements
NT2 international agreements
NT2 trade preferences
NT1 trade liberalization
NT1 trade negotiations
NT2 gatt
NT2 unctad
NT1 trade relations
rt policy

TRANSPORT
NT1 air transport
NT1 collection
NT1 container transport
NT1 inland transport
NT1 inland waterways
NT1 international transport
NT1 rail transport
NT1 road transport
NT1 sea transport
rt handling

VALUATION
NT1 ground rent
NT1 land prices
NT1 rent
NT1 taxable rent value
rt. compensation

VETERINARY PRODUCTS
rt veterinary services

WATER
rt irrigation
rt water management
rt water pollution
rt waterways

WILD LIFE
NT1 natural fauna
NT1 natural flora

YIELDS
- **rt** high yielding varieties
- **rt** productivity

AFRICA
NT1 africa south of sahara
NT2 central africa
NT3 burundi
NT3 cameroon
NT3 central african empire
NT3 chad
NT3 congo
NT3 equatorial guinea
NT3 gabon
NT3 rwanda
NT3 sao tome and principe
NT3 zaire
NT2 east africa
NT3 afars and issas territory
NT3 burundi
NT3 ethiopia
NT3 kenya
NT3 malawi
NT3 rwanda
NT3 seychelles
NT3 somalia
NT3 sudan
NT3 tanzania
NT3 uganda
NT2 southern africa
NT3 angola
NT3 botswana
NT3 comoro islands
NT3 lesotho
NT3 madagascar
NT3 malawi
NT3 mauritius
NT3 mozambique
NT3 namibia
NT3 reunion
NT3 rhodesia
NT3 south africa
NT3 st helena
NT3 swaziland
NT3 zambia
NT2 west africa
NT3 benin
NT3 cameroon
NT3 cape verde
NT3 gambia
NT3 ghana
NT3 guinea
NT3 guinea-bissau
NT3 ivory coast
NT3 liberia
NT3 mali
NT3 mauritania
NT3 niger
NT3 nigeria
NT3 senegal
NT3 sierra leone
NT3 st helena
NT3 togo
NT3 upper volta
NT3 western sahara
NT1 north africa
NT2 algeria
NT2 egypt
NT2 libya
NT2 morocco
NT2 tunisia
rt english-speaking africa
rt french-speaking africa

AMERICA
NT1 caribbean
NT2 bahamas
NT2 barbados
NT2 cayman islands
NT2 cuba
NT2 dominican republic
NT2 guadeloupe
NT2 haiti
NT2 jamaica
NT2 martinique
NT2 montserrat
NT2 netherlands antilles
NT2 puerto rico

AMERICA (C)
NT2 tortola
NT2 trinidad and tobago
NT2 turks and caicos islands
NT2 united states virgin islands
NT2 west indies associated states
NT3 antigua
NT3 dominica
NT3 grenada
NT3 st kitts and nevis and anguilla
NT3 st lucia
NT3 st vincent
NT1 central america
NT2 belize
NT2 costa rica
NT2 cuba
NT2 dominican republic
NT2 el salvador
NT2 guatemala
NT2 haiti
NT2 honduras
NT2 jamaica
NT2 nicaragua
NT2 panama
NT2 panama canal zone
NT2 puerto rico
NT1 north america
NT2 bermuda
NT2 canada
NT2 greenland
NT2 mexico
NT2 st pierre and miquelon
NT2 usa
NT1 south america
NT2 argentina
NT2 bolivia
NT2 brazil
NT2 chile
NT2 colombia
NT2 ecuador
NT2 falkland islands
NT2 french guiana
NT2 guyana
NT2 paraguay
NT2 peru
NT2 surinam
NT2 uruguay
NT2 venezuela

ANTARCTICA
rt australia
rt france
rt norway
rt united kingdom

ARAB COUNTRIES
NT1 algeria
NT1 egypt
NT1 iraq
NT1 jordan
NT1 lebanon
NT1 libya
NT1 morocco
NT1 saudi arabia
NT1 sudan
NT1 syria
NT1 tunisia
NT1 yemen
NT1 yemen pdr
rt north africa

ASIA
NT1 east asia
NT2 china
NT2 hong kong
NT2 japan
NT2 korea dpr
NT2 korea r
NT2 macao
NT2 mongolia
NT1 persian gulf states
NT2 bahrain
NT2 kuwait
NT2 qatar

ASIA (C)
NT2 united arab emirates
NT1 south east asia
NT2 bangladesh
NT2 bhutan
NT2 brunei
NT2 burma
NT2 india
NT2 indochina
NT3 kampuchea
NT3 lao
NT3 vietnam
NT2 indonesia
NT2 malaysia
NT2 maldives
NT2 nepal
NT2 pakistan
NT2 philippines
NT2 portuguese timor
NT2 singapore
NT2 sri lanka
NT2 thailand
NT1 west asia
NT2 afghanistan
NT2 cyprus
NT2 iran
NT2 iraq
NT2 israel
NT2 jordan
NT2 lebanon
NT2 oman
NT2 saudi arabia
NT2 syria
NT2 yemen
NT2 yemen pdr

COMMONWEALTH
rt australia
rt bahamas
rt bangladesh
rt barbados
rt botswana
rt canada
rt cyprus
rt fiji
rt gambia
rt ghana
rt guyana
rt india
rt jamaica
rt kenya
rt lesotho
rt malawi
rt malaysia
rt malta
rt mauritius
rt nauru
rt new zealand
rt nigeria
rt papua new guinea
rt pomegranates
rt sierra leone
rt singapore
rt sri lanka
rt swaziland
rt tonga
rt trinidad and tobago
rt uganda
rt united kingdom
rt western samoa
rt zambia

ENGLISH-SPEAKING AFRICA
rt africa
rt botswana
rt egypt
rt french-speaking africa
rt gambia
rt ghana
rt kenya
rt lesotho
rt liberia
rt libya
rt malawi

ENGLISH-SPEAKING AFRICA (C)
rt mauritius
rt namibia
rt nigeria
rt rhodesia
rt sierra leone
rt south africa
rt st helena
rt sudan
rt swaziland
rt tanzania
rt uganda
rt zambia

EUROPE
NT1 eastern europe
NT2 albania
NT2 bulgaria
NT2 byelorussian ssr
NT2 czechoslovakia
NT2 germany dr
NT2 hungary
NT2 poland
NT2 romania
NT2 ukrainian ssr
NT2 ussr
NT2 yugoslavia
NT1 western europe
NT2 andorra
NT2 austria
NT2 azores
NT2 belgium
NT3 belgium (antwerp)
NT3 belgium (brabant)
NT3 belgium (east flanders)
NT3 belgium (hainault)
NT3 belgium (liege)
NT3 belgium (limbourg)
NT3 belgium (namur)
NT3 belgium (west flanders)
NT2 canary islands
NT2 cyprus
NT2 france
NT3 france (overseas depts and terr)
NT3 france (r. alsace)
NT3 france (r. aquitaine)
NT3 france (r. auvergne)
NT3 france (r. brittany)
NT3 france (r. burgundy)
NT3 france (r. centre)
NT3 france (r. champagne ardennes)
NT3 france (r. corsica)
NT3 france (r. franche comte)
NT3 france (r. ile de france)
NT3 france (r. languedoc roussillon)
NT3 france (r. limousin)
NT3 france (r. lorraine)
NT3 france (r. lower normandy)
NT3 france (r. midi pyrenees)
NT3 france (r. nord pas de calais)
NT3 france (r. pays de la loire)
NT3 france (r. picardy)
NT3 france (r. poitou charentes)
NT3 france (r. provence alpes cote d'az
NT3 france (r. rhone alpes)
NT3 france (r. upper normandy)
NT2 germany fr
NT3 germany fr (baden-wuerttemberg)
NT3 germany fr (bavaria)
NT3 germany fr (hesse)
NT3 germany fr (lower saxony)
NT3 germany fr (north rhine westphalia)
NT3 germany fr (rhineland palatinate)
NT3 germany fr (saarland)
NT3 germany fr (schleswig-holstein)
NT2 gibraltar
NT2 greece
NT2 ireland
NT3 ireland (connacht)
NT3 ireland (leinster)
NT3 ireland (munster)
NT2 italy
NT3 italy (abruzzi)
NT3 italy (apulia)

EUROPE (C)
NT3 italy (basilicata)
NT3 italy (calabria)
NT3 italy (campania)
NT3 italy (emilia-romagna)
NT3 italy (friuli-venezia giulia)
NT3 italy (latium)
NT3 italy (liguria)
NT3 italy (lombardy)
NT3 italy (marches)
NT3 italy (molise)
NT3 italy (piedmont)
NT3 italy (sardinia)
NT3 italy (sicily)
NT3 italy (trentino-alto adige)
NT3 italy (tuscany)
NT3 italy (umbria)
NT3 italy (valle d'aosta)
NT3 italy (veneto)
NT2 liechtenstein
NT2 luxembourg
NT2 madeira
NT2 malta
NT2 monaco
NT2 netherlands
NT3 netherlands (drenthe)
NT3 netherlands (friesland)
NT3 netherlands (gelderland)
NT3 netherlands (groningen)
NT3 netherlands (n.brabant)
NT3 netherlands (n.holland)
NT3 netherlands (overijssel)
NT3 netherlands (s.holland)
NT3 netherlands (s.limburg)
NT3 netherlands (utrecht)
NT3 netherlands (zeeland)
NT2 portugal
NT2 san marino
NT2 scandinavia
NT3 denmark
NT4 denmark (bornholm)
NT4 denmark (fyn)
NT4 denmark (jutland)
NT4 denmark (lolland-falster)
NT4 denmark (zealand)
NT3 faroe islands
NT3 finland
NT3 greenland
NT3 iceland
NT3 norway
NT3 sweden
NT2 spain
NT2 switzerland
NT2 turkey
NT2 united kingdom
NT3 united kingdom (channel islands)
NT3 united kingdom (england and wales)
NT4 united kingdom (e.england)
NT4 united kingdom (e.midlands)
NT4 united kingdom (n.england)
NT4 united kingdom (s.w.england)
NT4 united kingdom (w.midlands)
NT4 united kingdom (wales)
NT4 united kingdom (yorks/lancs)
NT3 united kingdom (isle of man)
NT3 united kingdom (n.ireland)
NT3 united kingdom (scotland)
NT4 united kingdom (e.scotland)
NT4 united kingdom (highlands+islands)
NT4 united kingdom (n.scotland)
NT4 united kingdom (w.scotland)
NT2 vatican
rt european communities

FRENCH-SPEAKING AFRICA
rt afars and issas territory
rt africa
rt algeria
rt benin
rt burundi
rt cameroon
rt central african empire
rt chad
rt comoro islands

FRENCH-SPEAKING AFRICA (C)
rt congo
rt english-speaking africa
rt gabon
rt guinea
rt ivory coast
rt madagascar
rt mali
rt mauritania
rt morocco
rt niger
rt reunion
rt rwanda
rt senegal
rt togo
rt tunisia
rt upper volta
rt zaire

MEDITERRANEAN COUNTRIES
NT1 albania
NT1 algeria
NT1 cyprus
NT1 egypt
NT1 france
NT2 france (overseas depts and terr)
NT2 france (r. alsace)
NT2 france (r. aquitaine)
NT2 france (r. auvergne)
NT2 france (r. brittany)
NT2 france (r. burgundy)
NT2 france (r. centre)
NT2 france (r. champagne ardennes)
NT2 france (r. corsica)
NT2 france (r. franche comte)
NT2 france (r. ile de france)
NT2 france (r. languedoc roussillon)
NT2 france (r. limousin)
NT2 france (r. lorraine)
NT2 france (r. lower normandy)
NT2 france (r. midi pyrenees)
NT2 france (r. nord pas de calais)
NT2 france (r. pays de la loire)
NT2 france (r. picardy)
NT2 france (r. poitou charentes)
NT2 france (r. provence alpes cote d'az
NT2 france (r. rhone alpes)
NT2 france (r. upper normandy)
NT1 gibraltar
NT1 greece
NT1 israel
NT1 italy
NT2 italy (abruzzi)
NT2 italy (apulia)
NT2 italy (basilicata)
NT2 italy (calabria)
NT2 italy (campania)
NT2 italy (emilia-romagna)
NT2 italy (friuli-venezia giulia)
NT2 italy (latium)
NT2 italy (liguria)
NT2 italy (lombardy)
NT2 italy (marches)
NT2 italy (molise)
NT2 italy (piedmont)
NT2 italy (sardinia)
NT2 italy (sicily)
NT2 italy (trentino-alto adige)
NT2 italy (tuscany)
NT2 italy (umbria)
NT2 italy (valle d'aosta)
NT2 italy (veneto)
NT1 lebanon
NT1 libya
NT1 malta
NT1 monaco
NT1 morocco
NT1 san marino
NT1 spain
NT1 syria
NT1 tunisia
NT1 turkey
NT1 yugoslavia

OCEANIA
NT1 australia
NT1 fiji
NT1 nauru
NT1 new hebrides
NT1 new zealand
NT1 pacific islands au
NT1 pacific islands fr
NT2 french polynesia
NT2 new caledonia
NT1 pacific islands gb
NT1 pacific islands nz
NT1 pacific islands us
NT1 papua new guinea
NT1 tonga
NT1 western samoa

ORGANIZATIONS
NT1 acp
NT1 cacm
NT1 cmea
NT1 efta
NT1 european communities
NT1 european parliament
NT1 fao
NT1 lafta
NT1 oecd
NT1 un
NT1 unesco
NT1 who
NT1 world bank
rt interest groups
rt membership

ADDITIVES
NT1 antibiotics
NT1 colourings
NT1 flavourings
NT1 mineral supplements
NT1 preservatives
NT1 vitamins
rt feedingstuffs
rt food
rt manufactured feeds
rt preservation

ALCOHOL
rt alcoholic beverages

BAKERY PRODUCTS
NT1 bread
rt bakery industry
rt cereal products
rt food

BEES
rt apiculture
rt honey

BEVERAGES
NT1 alcoholic beverages
NT2 beer
NT2 cider
NT2 liqueurs
NT2 spirits
NT2 wine
NT3 champagne and sparkling wines
NT3 quality wine
NT3 table wine
NT1 soft drinks
NT2 coffee
NT2 mate
NT2 tea
rt food

CATTLE
NT1 beef cattle
NT1 buffaloes
NT1 bullocks
NT1 bulls
NT1 calves
NT1 cows
NT2 cows for beef
NT2 dairy cattle
NT2 nurse cows
NT1 draught animals
NT1 dual purpose cattle
NT1 heifers
rt cattle farming
rt livestock

CEREAL PRODUCTS
NT1 cornflakes
NT1 flour
NT2 cornflour
NT2 oatmeal
NT2 rice flour
NT2 rye flour
NT2 wheat flour
NT1 pasta
NT1 pearl barley
NT1 porridge oats
rt bakery products
rt cereals
rt food

CHOCOLATE
rt food
rt sugar confectionery

COCOA
rt food
rt instant cocoa

CROPS
NT1 catch crops
NT1 cereals
NT2 barley

CROPS (C)
NT3 barley for feed
NT3 malting barley
NT2 buckwheat
NT2 feed grains
NT3 coarse grains
NT2 food grains
NT2 maize
NT3 grain maize
NT3 silo maize
NT2 millet
NT2 oats
NT2 rice
NT2 rye
NT2 sorghum
NT2 triticale
NT2 wheat
NT3 hard wheat
NT3 soft wheat
NT1 culinary herbs
NT1 feed crops
NT2 clover
NT2 lucerne
NT1 fibre crops
NT2 cotton
NT2 fibre flax
NT2 hemp
NT2 kapok
NT2 sisal
NT1 grass
NT1 hops
NT1 medicinal herbs
NT1 oilseeds
NT2 castor beans
NT2 groundnuts
NT2 linseed
NT2 mustard
NT2 rape
NT2 sesame
NT2 soybeans
NT1 ornamental plants
NT2 flowers
NT3 bulbs
NT1 root crops
NT2 fodder beet
NT2 manioc
NT2 potatoes
NT3 feed potatoes
NT3 industrial potatoes
NT3 ware potatoes
NT2 sugar beet
NT2 turnips
NT2 yams
NT1 spice plants
NT1 sugar cane
NT1 trees and shrubs
rt cropping systems
rt field crops

EGG PRODUCTS
rt eggs
rt food

EGGS
rt egg products
rt food

FISH
rt fish farming
rt fish products

FISH PRODUCTS
rt fish
rt food
rt prepared foods

FRUIT PRODUCTS
NT1 dried fruit
NT2 prunes
NT2 raisins
NT1 fruit juices
NT1 jams
NT1 marmalades
NT1 nut products

FRUIT PRODUCTS (C)
rt food
rt prepared foods

FRUITS
NT1 berries
NT2 bilberries
NT2 blackberries
NT2 blackcurrants
NT2 gooseberries
NT2 raspberries
NT2 redcurrants
NT2 strawberries
NT1 citrus fruits
NT2 clementines
NT2 grapefruits
NT2 lemons
NT2 mandarins
NT2 oranges
NT1 figs
NT1 fresh fruits
NT1 grapes
NT1 melons
NT1 persimmons
NT1 pome fruits
NT2 apples
NT2 pears
NT2 quinces
NT1 pomegranates
NT1 rhubarb
NT1 stone fruits
NT2 apricots
NT2 cherries
NT2 olives
NT2 peaches
NT3 nectarines
NT2 plums
NT1 tropical fruits
NT2 avocadoes
NT2 bananas
NT2 dates
NT2 guavas
NT2 mangoes
NT2 passion fruit
NT2 pawpaws
NT2 pineapples
NT1 watermelons
NT1 wild fruits
rt dried fruit
rt nuts

GAME
NT1 deer
rt game birds
rt hunting

GOATS
NT1 billy goats
NT1 kids
NT1 nanny goats
rt goat keeping
rt livestock

HONEY
rt apiculture
rt bees
rt food

HORSES
NT1 foals
NT1 geldings
NT1 mares
NT1 stallions
rt horse breeding

INDUSTRIAL CROPS
NT1 rubber
NT1 tobacco
rt fibre crops
rt oilseeds

LIVESTOCK
rt cattle
rt goats

LIVESTOCK (C)
rt livestock numbers
rt pigs
rt poultry
rt sheep

MEAT
NT1 beef
NT1 horse meat
NT1 lamb
NT1 mutton
NT1 offals
NT1 pigmeat
NT1 poultry meat
NT1 veal
rt food
rt meat and livestock industry
rt meat products

MEAT PRODUCTS
NT1 bacon
NT1 dried meats
NT1 ham
NT1 sausages
rt food
rt meat
rt prepared foods

MILK
NT1 skim milk
rt food
rt milk products

MILK PRODUCTS
NT1 butter
NT1 cheese
NT2 hard cheese
NT2 processed cheese
NT2 soft cheese
NT1 cream
NT1 cultured milks
NT2 buttermilk
NT2 yoghurt
NT1 dried milk
NT2 dried skim milk
NT1 evaporated milk
NT1 flavoured milks
NT1 milk fat
NT1 milk proteins
NT1 sweetened condensed milk
NT1 whey
rt food
rt milk

NUTS
NT1 almonds
NT2 bitter almonds
NT2 sweet almonds
NT1 brazil nuts
NT1 cashew nuts
NT1 chestnuts
NT1 coconuts
NT1 cola nuts
NT1 hazelnuts
NT1 hickory nuts
NT1 pecans
NT1 pistachios
NT1 shea nuts
NT1 walnuts
rt fruits

OILS AND FATS
NT1 animal fats
NT1 fat products
NT1 vegetable oils
NT2 coconut oils
NT2 corn oils
NT2 cottonseed oils
NT2 groundnut oils
NT2 olive oil
NT2 palm oil
NT2 soy oils
NT2 sunflower seed oils
rt oils and fats industry

PIGS
NT1 boars
NT1 piglets
NT1 sows
rt livestock
rt pig farming

POULTRY
NT1 broilers
NT2 cockerels
NT2 day old chicks
NT2 hens
NT1 ducks
NT1 game birds
NT2 partridges
NT2 pheasants
NT2 pigeons
NT1 geese
NT1 turkeys
rt livestock
rt poultry farming

PREPARED FOODS
(foods which are more or less ready to consume. with or without heating or addition of liquids)
NT1 dietetic foods
NT2 infant foods
NT1 dried foods
NT2 freeze-dried foods
NT1 instant foods
NT2 instant cocoa
NT2 instant coffee
NT2 instant tea
rt fish products
rt food
rt fruit products
rt meat products
rt vegetable products

PRODUCTS
NT1 agricultural products
NT1 by-products
NT1 commodities
NT1 food
NT1 primary products
NT2 forest products
NT1 processed products

PROTEINS
NT1 animal proteins
NT1 synthetic proteins
NT1 vegetable proteins
rt nutrition

RABBITS
rt small animal rearing

SHEEP
NT1 dual purpose sheep
NT1 ewes
NT1 lambs
NT1 rams
NT1 wool sheep breeds
rt livestock
rt sheep farming

SHELLFISH
rt aquaculture

SIMULATED FOODS
NT1 artificial sweeteners
NT1 margarine
NT1 meat analogues
NT1 milk substitutes
rt substitution

SUGAR
NT1 beet sugar
NT1 cane sugar
NT1 isomerose
rt artificial sweeteners
rt food
rt sugar beet
rt sugar cane

SUGAR (C)
rt sugar confectionery

SUGAR CONFECTIONERY
rt chocolate
rt food
rt sugar

VEGETABLE PRODUCTS
NT1 potato products
NT1 soy products
rt food
rt prepared foods
rt vegetables

VEGETABLES
NT1 artichokes
NT1 asparagus
NT1 aubergines
NT1 beet
NT1 broccoli
NT1 brussels sprouts
NT1 cabbages
NT2 green cabbages
NT2 red cabbages
NT2 savoy cabbages
NT2 white cabbages
NT1 carrots
NT1 cauliflowers
NT1 celeriac
NT1 celery
NT1 chick peas
NT1 chicory
NT1 cucumbers
NT1 edible fungi
NT1 endives
NT1 fresh vegetables
NT1 garlic
NT1 gherkins
NT1 kohlrabi
NT1 leeks
NT1 legumes
NT2 beans
NT3 broad beans
NT3 french beans
NT3 horse beans
NT3 kidney beans
NT3 lima beans
NT3 white beans
NT2 lentils
NT2 peas
NT3 forage peas
NT1 lettuces
NT1 marrows
NT1 okra
NT1 onions
NT1 parsnips
NT1 peppers
NT1 pumpkins
NT1 radishes
NT1 shallots
NT1 spinach
NT1 squashes
NT1 swedes
NT1 sweet corn
NT1 sweet potatoes
NT1 tomatoes
NT1 watercress
rt vegetable growing
rt vegetable products

WILD LIFE
NT1 natural fauna
NT1 natural flora

AUTOMATION
rt computers
rt cybernetics
rt mechanization
rt technical progress

CENSUSES
NT1 agricultural censuses
NT1 population censuses
rt data collection
rt statistics
rt surveys

COMPUTER SOFTWARE
rt computers
rt electronic data processing

CONGRESSES
rt publications
rt trade associations

CYBERNETICS
rt automation
rt mathematics
rt operations research

DATA COLLECTION
rt censuses
rt data processing
rt sampling
rt surveys

DATA PROCESSING
NT1 on line
rt data collection

DOCUMENTATION
NT1 data banks
NT1 data bases
rt diffusion of research
rt history
rt information diffusion
rt information retrieval
rt rural sociology research

ECONOMETRICS
NT1 non-parametric methods
NT1 parametric methods
NT2 simultaneous equation analysis
NT2 single equation analysis
NT3 consumption functions
NT3 demand functions
NT3 elasticities
NT3 flexibility
NT3 investment functions
NT3 production functions
NT3 savings functions
NT3 supply functions
NT3 utility functions
rt operations research
rt sociological method

ELECTRONIC DATA PROCESSING
rt boolean algebra
rt computer software

EVALUATION
NT1 measurement
rt efficiency
rt planning

FORECASTS
NT1 projections

HARDWARE
NT1 computers
NT2 minicomputers

HISTORY
rt documentation

INFORMATION DIFFUSION
NT1 public relations
rt communication

INFORMATION DIFFUSION (C)
rt diffusion of research
rt documentation
rt extension
rt information processing
rt mass media

INFORMATION PROCESSING
NT1 document analysis
NT2 abstracting
NT2 classification
NT2 indexing
NT1 documentation languages
NT2 thesauri
NT1 information storage
NT2 microform
rt information diffusion

INFORMATION RETRIEVAL
NT1 card services
NT1 retrospective searches
NT1 sdi services
rt documentation

INFORMATION SERVICES
rt libraries

LIBRARIES
rt information services

MATHEMATICS
NT1 boolean algebra
NT1 calculus
NT1 financial mathematics
NT2 actuarial mathematics
NT1 matrices
NT1 modern mathematics
rt cybernetics
rt systems analysis
rt theory of graphs

MODELS
NT1 dynamic models
NT1 econometric models
NT1 farm models
NT1 international trade models
NT1 transport models

OPERATIONS RESEARCH
NT1 bayesian theory
NT1 non-optimizing methods
NT2 cost benefit analysis
NT2 farm budgeting
NT2 game theory
NT2 input-output analysis
NT2 markov chains
NT2 ppbs
NT2 simulation
NT3 monte carlo method
NT2 systems analysis
NT1 optimization methods
NT2 international trade models
NT2 transport models
NT1 plan implementation and evaluation
NT2 cash flow analysis
NT2 critical path analysis
NT2 network analysis
NT2 pert
NT2 queueing theory
NT1 project control
NT2 altai programme
NT2 budgetary control
rt cybernetics
rt econometrics
rt planning
rt sociological method
rt theory of graphs

PLANNING
NT1 agricultural planning
NT1 development plans
NT2 development projects
NT1 indicative planning
NT1 interregional planning

PLANNING (C)
NT1 intersectoral planning
NT1 local planning
NT1 market planning
NT1 national planning
NT1 normative planning
NT1 physical planning
NT2 open spaces
NT2 zoning
NT1 regional planning
NT1 rural planning
NT2 building controls
NT2 central places
NT1 sectoral planning
rt evaluation
rt farm planning
rt non-optimizing methods
rt operations research
rt optimization methods
rt plan implementation and evaluation
rt work planning

PROGRAMMING
NT1 computer programming
NT1 concave programming
NT1 convex programming
NT1 dynamic programming
NT1 integer programming
NT1 linear programming
NT1 non-linear programming
NT1 parametric programming
NT1 quadratic programming
NT1 recursive programming
NT1 stochastic programming
rt non-optimizing methods
rt optimization methods
rt project control

PUBLICATIONS
NT1 books
NT1 dictionary
NT1 directories
NT1 grey literature
NT1 maps
NT1 periodicals
NT2 secondary journals
NT3 bibliographies
NT1 reports
NT1 thesis
rt congresses

REMOTE SENSING
NT1 satellite surveys

STATISTICS
NT1 descriptive statistics
NT2 central value characteristics
NT3 averages
NT2 dispersion and distribution
NT3 standard deviation
NT3 time series
NT4 seasonality
NT3 variance
NT1 graphs
NT1 statistical analysis
NT2 sampling
NT2 statistical inference
NT3 chi square test
NT3 component analysis
NT3 correlation analysis
NT3 cross sectional analysis
NT3 discriminant analysis
NT3 factorial analysis
NT3 least squares method
NT3 multicollinearity
NT3 regression analysis
NT3 spectral analysis
NT3 student test
NT3 variance analysis
NT1 stochastic processes
NT2 probability analysis
NT1 theory of graphs
rt censuses

SURVEYS
NT1 farm surveys
NT2 european farm accounting network
rt censuses
rt data collection
rt sampling

AGRICULTURAL SCIENCES
rt agricultural research

CURRICULUM
rt education
rt literacy

EDUCATION
NT1 agricultural education
NT1 continuing education
NT1 literacy
NT1 schools
NT2 primary education
NT2 secondary education
NT1 specialized higher education
NT1 universities
NT1 vocational training
NT2 apprenticeship
NT2 on farm training
NT2 practical education
NT2 retraining
rt curriculum
rt leisure
rt teachers

EDUCATIONAL POLICY
NT1 apprenticeship articles
NT1 educational grants
NT1 educational reform
NT1 study leave
NT1 training levies
rt policy

EXTENSION
NT1 advisory centres
NT1 advisory officers
rt agricultural press
rt communication
rt diffusion of research
rt information diffusion
rt mass media

ORGANIZATION OF RESEARCH
NT1 research institutions
NT2 experimental stations
NT2 laboratories
NT2 pilot farms
NT2 research associations
NT2 research institutes
NT1 research personnel
NT2 research teams
NT2 research workers
NT3 agricultural economists
NT3 rural sociologists

RESEARCH
NT1 agricultural research
NT1 applied research
NT1 basic research
NT1 government research
NT1 private research
NT1 university research

RESEARCH POLICY
NT1 diffusion of research
NT1 evaluation of research
NT1 financing research
NT1 implementation of research
NT1 planning of research
NT2 research projects
NT1 research contracts
rt policy

STUDENTS
rt teachers

TEACHING METHODS
NT1 audiovisual aids
NT2 educational radio
NT2 educational television
NT1 case studies
NT1 discussion groups
NT1 management games
NT1 programmed learning
NT1 seminars

AGE
NT1 adults
NT1 age differences
NT1 age groups
NT1 age structure
NT2 population pyramids
NT1 childhood
NT1 generations
NT1 old age
NT1 youth
NT2 rural youth

AGRICULTURAL MANPOWER
NT1 child labour
NT1 elderly family-workers
NT1 family labour
NT1 hired labour
NT2 female labour
NT2 migrant labour
NT2 part time labour
NT2 permanent labour
NT2 seasonal labour
NT2 skilled labour
NT2 unskilled labour
NT1 manpower units
NT1 young workers
rt agricultural work force
rt farm workers
rt labour
rt labour economics
rt working population

BEHAVIOUR
NT1 attitudes
NT2 political attitudes
NT1 collective behaviour
NT1 consciousness
NT1 ideology
NT1 opinions
NT1 stereotypes
rt consumer behaviour
rt farmers attitudes

CHANGE
rt population change
rt social change
rt trends

COMMUNES
rt co-operation
rt collective farms
rt groups

COMMUNICATION
rt extension
rt information diffusion
rt mass media

CULTURE
NT1 civilization
NT1 cultural development
NT1 cultural environment
NT2 acculturation
NT2 cultural differentiation
NT2 cultural integration
NT1 cultural heritage
NT1 value systems
NT1 values
rt social anthropology

DECISION MAKING
NT1 action
NT1 cognition
NT1 motivation
NT1 thinking
rt bayesian theory

DEMOGRAPHY
rt population

ECONOMIC SOCIOLOGY
NT1 consumer behaviour
NT2 eating habits
NT2 purchasing habits

ECONOMIC SOCIOLOGY (C)
NT1 farmers attitudes
rt consumer expenditure
rt consumer motivation

EMPLOYMENT
NT1 occupational change
NT1 off-farm employment
NT1 underemployment
NT1 unemployment
NT2 seasonal unemployment
NT2 structural unemployment
rt labour
rt labour economics

ENVIRONMENT
NT1 ecology
NT1 natural disasters
NT1 rural environment
NT2 agricultural geography
NT2 rural areas
NT2 rural communities
NT3 villages
NT2 rural development
NT3 community development
NT4 rural animation
NT3 settlement
NT1 rural-urban relations
NT2 urbanization
NT1 urban environment
NT2 urban areas
NT3 suburban agriculture
NT3 towns
NT3 urban hinterland
rt environmental policy
rt landscape
rt pollution by agriculture
rt pollution of agriculture

FAMILIES
NT1 family and environment
NT1 family size
NT2 children
NT3 daughters
NT3 sons
NT2 grandparents
NT2 kinship
NT2 parents
NT3 fathers
NT3 mothers
NT2 youth
NT3 rural youth
NT1 farm families
NT1 marriage
NT2 divorce
NT2 marriage age
NT1 sexual behaviour
rt households
rt small groups

FAMILY LIFE
NT1 family breakdown
NT1 family integration
NT1 generation conflict
NT2 parent child relationships
NT1 woman's status

FAMILY STRUCTURE
NT1 extended families
NT2 multi-generation families
NT1 heads of families
NT1 matriarchy
NT1 nuclear families
NT2 one generation families
NT2 two generation families
NT1 partnerships
NT1 patriarchy
rt family size
rt structure

GROUPS
NT1 age groups
NT1 clubs
NT1 ethnic groups

GROUPS (C)
NT1 group integration
NT2 group behaviour
NT2 group consciousness
NT2 group interaction
NT2 group solidarity
NT1 group size
NT2 large groups
NT2 small groups
NT3 neighbourhood
NT1 interest groups
NT2 agricultural trade unions
NT2 farmers associations
NT2 pressure groups
NT2 trade associations
NT2 trade unions
NT1 membership
NT2 individuals and groups
NT1 working groups
rt communes

INDIVIDUALS
NT1 character
NT1 identity
NT1 interpersonal relations
NT1 self esteem
NT1 self perception

LEISURE
NT1 entertainment
NT1 sport
rt education
rt fishing
rt hunting
rt labour economics
rt mass media

LIVING CONDITIONS
NT1 living standards
NT2 basic needs
NT2 poverty
NT3 hunger
NT1 rural housing
NT2 farm dwellings
NT1 rural welfare
rt socioeconomic status

MASS MEDIA
NT1 films
NT1 newspapers
NT2 agricultural press
NT1 radio
NT1 television
rt communication
rt extension
rt information diffusion
rt leisure

MIGRATION
NT1 emigration
NT1 geographical mobility
NT2 commuting
NT2 rural-urban migration
NT2 seasonal migration
NT2 urban rural migration
NT1 immigration
NT1 rural depopulation

NATIONALISM
NT1 national character
NT1 national consciousness
rt nations
rt political systems

NATIONS
rt nationalism

POPULATION
NT1 population structure
NT2 agricultural population
NT2 non-gainfully employed
NT3 retired people
NT3 unemployed
NT2 rural population

POPULATION (C)
NT2 urban population
NT2 working population
NT3 agricultural work force
rt demography

POPULATION CHANGE
NT1 birth rate
NT1 family planning
NT1 health
NT1 mortality
NT1 population decrease
NT1 population density
NT1 population equilibrium
NT1 population forecasts
NT1 population growth
NT1 population pressure
rt change

RURAL ECONOMY
NT1 rural unemployment

RURAL SOCIOLOGY
rt rural-urban migration

RURAL SOCIOLOGY RESEARCH
NT1 sociological analysis
NT1 sociological method
NT1 sociology
rt documentation
rt research institutions
rt social geography

SOCIAL ANTHROPOLOGY
NT1 ethnic groups
rt culture

SOCIAL CHANGE
NT1 modernization
NT2 adoption of innovations
NT2 industrialization
NT2 technical progress
NT3 appropriate technology
NT3 technology transfers
NT1 social development
rt change
rt economic development

SOCIAL GEOGRAPHY
rt agricultural geography
rt geography
rt rural sociology research

SOCIAL STRUCTURE
NT1 social equilibrium
NT1 social stratification
NT2 ownership conditions
NT2 roles
NT3 aspirations
NT3 role changes
NT3 role conflicts
NT2 social classes
NT3 middle classes
NT3 ruling classes
NT3 upper classes
NT3 working classes
NT2 social mobility
NT3 horizontal mobility
NT3 vertical mobility
NT4 downward mobility
NT4 upward mobility
NT2 status
NT3 elites
NT3 occupational prestige
NT3 occupational status
NT3 socioeconomic status
rt structure

SOCIAL SYSTEMS
NT1 legal systems
NT2 legal rights
NT1 power
NT2 authority
NT2 leadership

SOCIAL SYSTEMS (C)
NT2 political power
NT1 religions
NT2 churches
NT3 churches in rural areas
NT1 social institutions
NT2 institution building
NT1 social sanctions
NT1 social standards
NT2 social customs
NT2 traditions

SOCIALIZATION
(sociology)
NT1 collectivization
NT1 social adjustment
NT2 maladjustment
NT1 social consciousness
NT1 social crisis
NT2 conflict
NT2 deviant behaviour
NT3 alcoholism
NT3 criminality
NT2 social isolation
NT2 social unrest
NT3 class conflict
NT1 social differentiation
NT1 social integration
NT1 social participation
NT2 social activity
rt co-operation

SOCIETIES
NT1 agricultural society
NT2 peasantry
NT1 industrial societies
NT1 mass society
NT1 pastoral society
NT2 nomads
NT1 plural societies
NT1 rural societies
NT1 urban society

SOCIOLOGY OF WORK
rt work study
rt working conditions

WOMEN
NT1 rural women
rt female labour
rt woman's status

WORKING CONDITIONS
NT1 attitudes to work
NT1 retirement
NT1 safety at work
NT2 accident prevention
NT2 industrial illness
NT1 work conflicts
NT1 work places
NT1 work satisfaction
NT1 work stress
NT1 working hours
rt labour economics
rt labour protection
rt sociology of work

Bei Fragen zur Produktsicherheit wenden Sie sich bitte an:
If you have any questions regarding product safety,
please contact:

Walter de Gruyter GmbH
Genthiner Straße 13
10785 Berlin
productsafety@degruyterbrill.com